# Parent and Student Study Guide Workbook

GLENCOE MATHEMATICS

# Mathematics

## Applications and Concepts

Course 1

McGraw Hill Glencoe

New York, New York   Columbus, Ohio   Chicago, Illinois   Peoria, Illinois   Woodland Hills, California

Glencoe

The McGraw·Hill Companies



Send all inquiries to:
The McGraw-Hill Companies
8787 Orion Place
Columbus, OH 43240-4027

ISBN: 0-07-860088-X

*Mathematics: Applications and Concepts, Course 1*
*Parent and Student Study Guide*

7 8 9 10 045 10 09 08 07 06 05

# Contents

## To The Parents of *Glencoe Mathematics* Students:

You teach your children all the time. You taught language to your infants and you read to your son or daughter. You taught them how to count and use basic arithmetic. Here are some ways you can continue to reinforce mathematics learning.

- Encourage a positive attitude toward mathematics.
- Set aside a place and a time for homework.
- Be sure your child understands the importance of mathematics achievement.

The *Glencoe* ***Parent and Student Study Guide Workbook*** is designed to help you support, monitor, and improve your child's math performance. These worksheets are written so that you do not have to be a mathematician to help your child.

The ***Parent and Student Study Guide Workbook*** includes:

- A 1-page *worksheet* for every lesson in the Student Edition (90 in all). Completing a worksheet with your child will reinforce the concepts and skills your child is learning in math class. Upside-down answers are provided right on the page.
- A 1-page *chapter review* for each chapter (14 in all). These worksheets review the skills and concepts needed for success on tests and quizzes. Answers are located on pages 105–108.

**Online Resources**
For your convenience, these worksheets are also available in a printable format at **msmath1.net/parent_student**.

**Online Study Tools** can help your student succeed.

- **msmath1.net/extra_examples**
  shows you additional worked-out examples that mimic the ones in the textbook.
- **msmath1.net/self_check_quiz**
  provides a self-checking practice quiz for each lesson.
- **msmath1.net/vocabulary_review**
  checks your understanding of the terms and definitions used in each chapter.
- **msmath1.net/chapter_test**
  allows you to take a self-checking test before the actual test.
- **msmath1.net/standardized_test**
  is another way to brush up on your standardized test-taking skills.

NAME ______________________ DATE ____________ PERIOD _____

# 1-1 A Plan for Problem Solving (pages 6–9)

You can use a four-step plan to solve a problem.

| | |
|---|---|
| **Explore** | Read the problem carefully. Ask yourself questions like, "What facts do I know?" |
| **Plan** | See how the facts relate to each other. Make a plan for solving the problem. Estimate the answer. |
| **Solve** | Use your plan to solve the problem. If your plan does not work, revise it or make a new one. |
| **Examine** | Reread the problem. Ask, "Is my answer close to my estimate and does my answer make sense?" If not, solve the problem another way. |

## EXAMPLE

Efrain wants to buy a used book that costs 99 cents. He has three quarters and four dimes in his pocket. Does he have enough money to buy the book?

| | |
|---|---|
| **Explore** | You need to find out if Efrain has enough money to buy the book. With the coins he has, you estimate that he has enough money. |
| **Plan** | Multiply the number of quarters he has by 25, and the number of dimes he has by 10. Add the two products to find out how much money he has. |
| **Solve** | $3 \times 25 + 4 \times 10 = 115$ cents, and $115 > 99$ |
| **Examine** | Since Efrain has 115 cents, or $1.15, he can buy the book. |

### Try This Together

1. Lawanda sells candy bars for $2 each. How many bars must she sell to raise $60? *HINT: What must you multiply by $2 to get a product of $60?*

## PRACTICE

***Use the four-step plan to solve each problem.***

2. Find the next three numbers in the pattern 2, 3, 5, 8, __?__, __?__, __?__.
3. **Food** Erika is making cookies. The recipe she has makes 20 cookies, but she wants to make 60 cookies. If she needs 2 cups of flour for 20 cookies, how many cups of flour will she need for 60 cookies?
4. **Standardized Test Practice** Miguel rode his bike to swimming practice and home again every day for 80 days over the summer. The ride was 3 miles to practice and 3 miles back home. Altogether, how many miles did Miguel ride his bike to and from swimming practice?

   **A** 560 miles **B** 240 miles **C** 480 miles **D** 125 miles

**Answers: 1.** 30 candy bars **2.** 12, 17, 23 **3.** 6 cups **4.** C

NAME ____________________ DATE __________ PERIOD ____

# 1-2 Divisibility Patterns (pages 10–13)

When you divide a whole number by another whole number, and the quotient is a whole number, then the first number is divisible by the second. For example, 12 is divisible by 2 because the quotient 12 ÷ 2 is 6. You can test for divisibility mentally by using the divisibility rules below.

| | |
|---|---|
| **Divisibility Rules for 2, 3, 4, 5, 6, 9, 10** | A number is divisible by:<br>• 2 if the ones digit is divisible by 2.<br>• 3 if the sum of the digits is divisible by 3.<br>• 4 if the number formed by the last two digits is divisible by 4.<br>• 5 if the ones digit is 0 or 5.<br>• 6 if the number is divisible by both 2 and 3.<br>• 9 if the sum of the digits is divisible by 9.<br>• 10 if the ones digit is 0. |

## EXAMPLES

**A** Is 34 divisible by 2?

*The ones digit is 4. Since 4 ÷ 2 = 2, 4 is divisible by 2. So, 34 is divisible by 2.*

**B** Is 52 divisible by 3?

*The sum of the digits is 5 + 2, or 7. Since 7 is not divisible by 3, 52 is not divisible by 3.*

### Try These Together

**1.** Is 70 divisible by 5?
*HINT: Is the ones digit 0 or 5?*

**2.** Is 208 divisible by 9?
*HINT: Is the sum of the digits divisible by 9?*

## PRACTICE

***Tell whether the first number is divisible by the second number.***

**3.** 984; 2
**4.** 533; 4
**5.** 935; 5
**6.** 570; 3
**7.** 2,861; 2
**8.** 626; 6
**9.** 5,650; 10
**10.** 8,844; 6
**11.** 77,787; 9

***Tell whether each number is divisible by 2, 3, 4, 5, 6, 9, or 10.***

**12.** 365
**13.** 1,170
**14.** 887
**15.** 486
**16.** 620
**17.** 2,865
**18.** 350
**19.** 4,544
**20.** 51

**21. Design** The fourth grade class at Chavez Elementary School is having a group photo taken. There are 102 students in the fourth grade. Can they form 6 equal rows for the photo?

**22. Standardized Test Practice** Which number is divisible by both 2 and 9?

**A** 5,148 **B** 5,618 **C** 8,364 **D** 9,782

**Answers:** 1. yes 2. no 3. yes 4. no 5. yes 6. yes 7. no 8. no 9. yes 10. yes 11. yes 12. 5 13. 2, 3, 5, 6, 9, 10 14. none 15. 2, 3, 6, 9 16. 2, 4, 5, 10 17. 3, 5 18. 2, 5, 10 19. 2, 4 20. 3 21. yes 22. A

NAME ______________________ DATE ____________ PERIOD _____

# 1-3 Prime Factors (pages 14–17)

A **composite number** is any whole number greater than one that has more than two factors.

A number with only 2 factors, 1 and the number itself, is a **prime number**. The numbers 0 and 1 are neither prime nor composite.

Every composite number can be expressed as a product of prime numbers. This is called the **prime factorization** of the number. You can use a **factor tree** to find prime factorizations.

## EXAMPLES

**A** Is 7 a prime number?

*How many rectangles can you make out of 7 squares?* $1 \times 7$

*Only one rectangle, so the factors of 7 are 1 and 7. Since there are only 2 factors, 7 is a prime number.*

**B** Find the prime factorization of 12.

*Use a factor tree.*

12

(2) × 6

(2) × (3)

*Factor 12. 12 is divisible by 2. Circle the prime number 2.*

*Factor 6. 6 is divisible by 2. Circle the prime numbers 2 and 3. The prime factorization is $2 \times 2 \times 3$.*

## Try These Together

**1.** Is 22 a prime number?
*HINT: Does it have more than 2 factors?*

**2.** Find the prime factorization of 18.
*HINT: Use a factor tree to find prime factors.*

## PRACTICE

***Tell whether each number is* prime, composite, *or* neither.**

**3.** 2 **4.** 11 **5.** 14
**6.** 1 **7.** 84 **8.** 31
**9.** 111 **10.** 0 **11.** 113

***Find the prime factorization of each number.***

**12.** 10 **13.** 33 **14.** 87
**15.** 54 **16.** 29 **17.** 34
**18.** 61 **19.** 57 **20.** 112

**21. Entertainment** A cable system has 42 channels. Express 42 as a product of primes.

**22. Standardized Test Practice** What is the least prime number greater than 50?

**A** 51 **B** 53 **C** 57 **D** 59

**Answers: 1.** no **2.** $2 \times 3 \times 3$ **3.** prime **4.** prime **5.** composite **6.** neither **7.** composite **8.** prime **9.** composite **10.** neither **11.** prime **12.** $2 \times 5$ **13.** $3 \times 11$ **14.** $3 \times 29$ **15.** $2 \times 3 \times 3 \times 3$ **16.** prime **17.** $2 \times 17$ **18.** prime **19.** $3 \times 19$ **20.** $2 \times 2 \times 2 \times 2 \times 7$ **21.** $2 \times 3 \times 7$ **22.** B

# 1-4 Powers and Exponents (pages 18–21)

When you multiply two or more numbers, each number is called a **factor** of the product. When the same factor is repeated, you can use an exponent to simplify your writing. An **exponent** tells you how many times a number, called the **base**, is used as a factor. A **power** is a number that is expressed using exponents.

| | |
|---|---|
| **Order of Operations with Powers** | 1. Do all powers before other operations.<br>2. Multiply and divide in order from left to right.<br>3. Add and subtract in order from left to right. |

## EXAMPLES

**A** Write $7 \cdot 7 \cdot 7$ using exponents.
*The base is 7. Since 7 is a factor three times, the exponent is 3.*
$7 \cdot 7 \cdot 7 = 7^3$

**B** Write $9^2$ as a product. Then find the value of the product.
*The base is 9. The exponent 2 means that 9 is a factor two times.*
$9^2 = 9 \cdot 9 = 81$

**C** Write the prime factorization of 54 using exponents.
*The prime factorization of 54 is $2 \times 3 \times 3 \times 3$, or $2 \times 3^3$.*

### Try These Together

**1.** Write $21 \cdot 21 \cdot 21$ using exponents.
*HINT: How many factors are there?*

**2.** Write $4^4$ as a product. Then find the value of the product.
*HINT: How many times is 4 a factor?*

## PRACTICE

***Write each product using an exponent. Then find the value of the power.***

**3.** $12 \cdot 12$ **4.** $5 \cdot 5 \cdot 5 \cdot 5$ **5.** $2 \cdot 2 \cdot 2 \cdot 2 \cdot 2$ **6.** $6 \cdot 6 \cdot 6$

***Write each power as a product. Then find the value of the product.***

**7.** $6^4$ **8.** $36^2$ **9.** $3^4$ **10.** $10^3$

***Write the prime factorization of each number using exponents.***

**11.** 63 **12.** 52 **13.** 28 **14.** 81

**15. Population** The U.S. Census Bureau estimated in 1999 that there were about $10^7$ 60 to 64-year-olds living in the United States. About how many people is this?

**16. Standardized Test Practice** Rewrite $2 \times 2 \times 3 \times 3 \times 7$ using exponents.

**A** $2^2 \times 3^2 \times 7$ **B** $2 \times 3^2 \times 7$ **C** $2^3 \times 3^2 \times 7$ **D** $2^2 \times 3 \times 7$

**Answers: 1.** $21^3$ **2.** $4 \times 4 \times 4 \times 4$; 256 **3.** $12^2$; 144 **4.** $5^4$; 625 **5.** $2^5$; 32 **6.** $6^3$; 216 **7.** $6 \cdot 6 \cdot 6 \cdot 6$; 1,296 **8.** $36 \cdot 36$; 1,296 **9.** $3 \cdot 3 \cdot 3 \cdot 3$; 81 **10.** $10 \cdot 10 \cdot 10$; 1,000 **11.** $3^2 \times 7$ **12.** $2^2 \times 13$ **13.** $2^2 \times 7$ **14.** $3^4$ **15.** 10,000,000 **16.** A

NAME ______________________ DATE ____________ PERIOD ______

# 1-5 Order of Operations (pages 24–27)

Pemdas

When you have more than one operation, the order of operations tells you which operation to use first.

| | |
|---|---|
| **Order of Operations** | 1. Simplify the expressions inside grouping symbols, like parentheses.<br>2. Find the value of all powers.<br>3. Multiply and divide in order from left to right.<br>4. Add and subtract in order from left to right. |

## EXAMPLES

Find the value of each expression.

**A** $25 - 2^2 \times 6$

*$25 - 2^2 \times 6 = 25 - 4 \times 6$ Find $2^2$.*
*$= 25 - 24$ Multiply 4 and 6.*
*$= 1$ Subtract 24 from 25.*

**B** $(2 + 10) \div 3$

*$(2 + 10) \div 3 = 12 \div 3$ Add 2 and 10.*
*$= 4$ Divide 12 by 3.*

### Try These Together

***Find the value of each expression.***

**1.** $8 - 5 + 13$ 16
*HINT: Add and subtract from left to right.*

**2.** $(3^2 + 7) \times 2$ 36
*HINT: Simplify within parentheses first.*

## PRACTICE

***Find the value of each expression.***

**3.** $10 - 5 + 3^3$ 14
**4.** $8 \times 2 - 16$ 0
**5.** $(15 - 3) \times 2$ 24
**6.** $(12 + 4) \times 3$ 48
**7.** $1 + (4 - 3) \times 2^3$ 9
**8.** $2^2 \times (3 - 1)$ 8
**9.** $5 \times (5^2 + 5)$ 150
**10.** $6 \times 10 - (40 + 2)$ 18
**11.** $24 \div 3 - 6$ 2
**12.** $50 \div 5 + 15$ 25
**13.** $27 \div 9 \times 4$ 12
**14.** $(18 - 3) \times 5$ 75

**15.** Find the value of $2^2 + 8 \times 3 - 6$.

**16.** What is the value of 10 times 3 divided by 6?

**17. Money Matters** Cassie makes \$2 for taking out the trash and \$1 for making her bed. If she took out the trash 3 times, and made her bed 2 times, how much money did she make? \$8

**18. Standardized Test Practice** Jackson had 10 baseball cards. He bought 10 more. Then he divided the cards evenly between 5 people. How many baseball cards did each person receive?

**A** 3 **B** 6 **C** 5 **D** 4

**Answers:** **1.** 16 **2.** 32 **3.** 32 **4.** 0 **5.** 24 **6.** 48 **7.** 9 **8.** 8 **9.** 150 **10.** 18 **11.** 2 **12.** 25 **13.** 12 **14.** 75 **15.** 22 **16.** 5 **17.** \$8 **18.** D

NAME ______________________ DATE ____________ PERIOD _____

# 1-6 Variables and Expressions (pages 28–31)

In algebra, **variables**, usually letters, are used to represent numbers. **Algebraic expressions** are combinations of variables, numbers, and at least one operation. If you replace variables with numbers, you can **evaluate**, or find the value of, an algebraic expression.

## EXAMPLES

***Evaluate each expression if h = 9.***

**A** $26 - h$

$26 - h = 26 - 9$ *Replace h with 9.*
$= 17$ *Subtract 9 from 26.*

**B** $4h + 8$

$4h + 8 = 4 \times 9 + 8$ *Replace h with 9.*
$= 36 + 8$ *Multiply 4 by 9.*
$= 44$ *Add 36 and 8.*

### Try These Together

***Evaluate each expression if q = 7 and r = 4.***

**1.** $q + r - 1$
*HINT: Replace the variables.*

**2.** $3q + r$
*HINT: Replace the variables, then multiply first.*

## PRACTICE

***Evaluate each expression if x = 4 and y = 9.***

**3.** $x + 7$ **4.** $18 - y$ **5.** $6x - 10$
**6.** $6 + y$ **7.** $2xy$ **8.** $y \div 1$
**9.** $x + 3x$ **10.** $x \times y$ **11.** $40 \div 5x$

***Evaluate each expression if a = 9, b = 18, and c = 3.***

**12.** $b \div 6$ **13.** $b - c$ **14.** $ca$
**15.** $a + b + c$ **16.** $ab - c$ **17.** $54 \div a$
**18.** $cb + 2a$ **19.** $b - 2a$ **20.** $b - 3a \div c$

**21. Architecture** To find the perimeter of a rectangle, you can use the expression $2\ell + 2w$ where $\ell$ and $w$ represent the length and width of the rectangle. Find the perimeter of a rectangle with length 4 m and width 7 m.

**22. Standardized Test Practice** Evaluate $15 - st$ if $s = 2$ and $t = 3$.

**A** 23 **B** 10 **C** 9 **D** 21

Answers: **1.** 10 **2.** 25 **3.** 11 **4.** 9 **5.** 14 **6.** 15 **7.** 72 **8.** 9 **9.** 16 **10.** 36 **11.** 2 **12.** 3 **13.** 15 **14.** 27 **15.** 30 **16.** 159 **17.** 6 **18.** 72 **19.** 0 **20.** 9 **21.** 22 m **22.** C

NAME ______________________________ DATE ____________ PERIOD _____

# 1-7 Solving Equations (pages 34–37)

In mathematics, an **equation** is a sentence that contains an equals sign, =. Equations can be either true or false. An equation with a variable is neither true nor false until the variable is replaced with a number.

| | |
|---|---|
| $y + 2 = 9$<br>Replace $y$ with 5.<br>Is $5 + 2 = 9$ a true sentence?<br>$7 = 9$<br>No, the sentence is false. | $y + 2 = 9$<br>Replace $y$ with 7.<br>Is $7 + 2 = 9$ a true sentence?<br>$9 = 9$<br>Yes, the sentence is true. The solution of $y + 2 = 9$ is 7. |

## EXAMPLES

**A** Is $12 - z = 10$ true if $z = 3$?

*$12 - 3 \stackrel{?}{=} 10$ Replace z with 3.*

*$9 = 10$ Subtract 3 from 12.*

*No, the sentence is false.*

**B** Is $3a + 1 = 13$ true if $a = 4$?

*$3 \cdot 4 + 1 \stackrel{?}{=} 13$ Replace a with 4.*

*$12 + 1 \stackrel{?}{=} 13$ Multiply 3 by 4.*

*$13 = 13$ Add 12 and 1.*

*Yes, the sentence is true. The solution of $3a + 1 = 13$ is 4.*

## Try These Together

***Identify the solution of each equation from the list given.***

**1.** $s + 15 = 19$; 3, 4, 5

*HINT: Replace the variable, then evaluate.*

**2.** $n - 7 = 2$; 7, 8, 9

*HINT: Replace the variable, then evaluate.*

## PRACTICE

***Tell whether the equation is true or false by replacing the variable with the given value.***

**3.** $75 + s = 120$; $s = 45$

**4.** $95 \div x = 5$; $x = 17$

**5.** $y - 22 = 56$; $y = 78$

**6.** $6m = 48$; $m = 7$

***Identify the solution of each equation from the list given.***

**7.** $j + 4 = 21$; 17, 18, 19

**8.** $b - 77 = 32$; 107, 109, 111

**9.** $45 = 15r$; 3, 4, 5

**10.** $27 + w = 45$; 17, 18, 19

***Solve each equation mentally.***

**11.** $6 + p = 14$

**12.** $75 = 3t$

**13.** $18v = 36$

**14.** **Standardized Test Practice** Solve $39 \div s = 3$.

**A** 3 **B** 6 **C** 11 **D** 13

Answers: **1.** 4 **2.** 9 **3.** true **4.** false **5.** true **6.** false **7.** 17 **8.** 109 **9.** 3 **10.** 18 **11.** 8 **12.** 25 **13.** 2 **14.** D

NAME ________________ DATE ________ PERIOD ______

# 1-8 Area of Rectangles (pages 39–41)

The **area** ($A$) of a closed figure is the number of square units needed to cover its surface. You can use algebra to help you find the area of a rectangle.

| | |
|---|---|
| **Area of a Rectangle** | The area of a rectangle is the product of its length $\ell$ and width $w$, or $A = \ell \cdot w$. 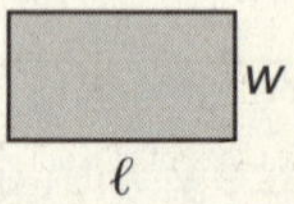 |



## EXAMPLES

**A** Find the area of a rectangle with a length of 9 cm and a width of 4 cm.

$A = \ell \cdot w$
$A = 9 \cdot 4$ — *$\ell = 9$ and $w = 4$*
$A = 36$ — *The area is 36 square centimeters.*

**B** Find the area of a rectangle with a length of 12 ft and a width of 6 ft.

$A = \ell \cdot w$
$A = 12 \cdot 6$ — *$\ell = 12$ and $w = 6$*
$A = 72$ — *The area is 72 square feet.*

### Try These Together

1. Find the area of a rectangle with a length of 8 yd and a width of 5 yd.
   *HINT: Area of a rectangle is length times width.*
2. Find the area of a rectangle with a length of 9 m and a width of 7 m.

## PRACTICE

***Find the area of each figure.***

3. 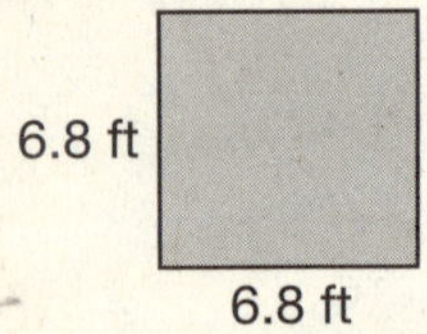


4. 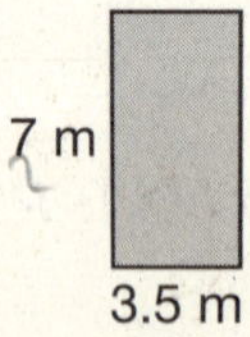


5. 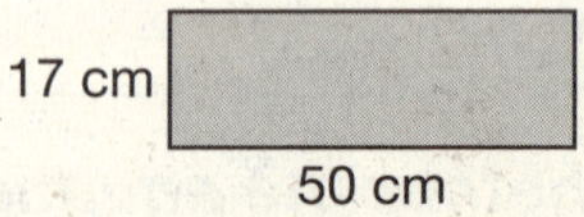


6. square: $s = 7.1$ in.
7. rectangle: $\ell = 33$ ft, $w = 70$ ft
8. square: $s = 6.2$ cm
9. square: $s = 12.5$ yd
10. rectangle: $\ell = 5$ m, $w = 9$ m
11. rectangle: $\ell = 24$ in., $w = 66$ in.
12. **Standardized Test Practice** A rectangle is 6 cm long, and its area is 18 $cm^2$. What is its width?

**A** 9 cm **B** 6 cm **C** 5 cm **D** 3 cm

**Answers: 1.** 40 $yd^2$ **2.** 63 $m^2$ **3.** 46.24 $ft^2$ **4.** 24.5 $m^2$ **5.** 850 $cm^2$ **6.** 50.41 $in^2$ **7.** 2310 $ft^2$ **8.** 38.44 $cm^2$ **9.** 156.25 $yd^2$ **10.** 45 $m^2$ **11.** 1,584 $in^2$ **12.** D

NAME ______________________ DATE __________ PERIOD ____

# 1 Chapter 1 Review

## Password Search

The Middle School Math Club has just started their Web site.
For fun, they put a password on their site.
You can find the password using the clues.

**Clue 1:** Write the second step in the four-step problem solving plan here. Plan
Write the first letter of this word in blank 1 in the box at the bottom of the page.

**Clue 2:** The sixth number of the following pattern.
71, 62, 53, 44, 35, 26

Find the value of each expression. Use the chart to translate each solution to a letter. Write the letter in the blank that matches the number of the clue.

**Clue 3:** $15 + (8 \div 2) \times 3 - 3$

**Clue 4:** $a^3 - 5b$ if $a = 3$ and $b = 5$

**Clue 5:** Use mental math to solve $42 \div w = 7$.

| Number | Letter | Number | Letter |
|---|---|---|---|
| 1 | X | 14 | U |
| 2 | E | 15 | L |
| 3 | C | 16 | I |
| 4 | A | 17 | D |
| 5 | Z | 18 | G |
| 6 | R | 19 | K |
| 7 | Y | 20 | N |
| 8 | S | 21 | Q |
| 9 | M | 22 | V |
| 10 | T | 23 | P |
| 11 | B | 24 | W |
| 12 | F | 25 | H |
| 13 | J | 26 | O |

What is the password?

**Password**

When you enter the Middle School Math Club Web site, you will gain math . . .

P O W E R .
1 2 3 4 5

Answers are located on p. 105.

NAME ______________________ DATE ____________ PERIOD _____

# 2-1 Frequency Tables (pages 50–53)

When you use **statistics**, you collect, organize, analyze, and present data, often as a **frequency table**.

| | |
|---|---|
| **Choosing a Scale for a Frequency Table** | Choose a scale that includes the least and the greatest number. • Choose an interval that will give you a manageable number of groups, usually from four to seven. • Make sure all the intervals, or groups, are equal and they do not overlap. |
| **Making a Frequency Table** | • Draw a table with three columns and tally the responses. In the third column, write the number of tallies (or frequency). |

## EXAMPLES

**A** Name the scale and the interval in this first column of a frequency table:

**Free Throws**
16–20
11–15
6–10
1–5

*The scale goes from 1 to 20. Each interval has 5 scores in it (for example, 16, 17, 18, 19, 20). The interval is 5.*

**B** Here are the number of free throws made by the third period gym class: 17, 2, 10, 4, 5, 7, 7, 16, 3, 12, 9, 3, 4. Complete the frequency table started in Example A.

*Add two columns to the table. Mark tallies for each interval. Then write the frequencies.*

| Free Throws | Tally | Frequency |
|---|---|---|
| 16–20 | II | 2 |
| 11–15 | I | 1 |
| 6–10 | IIII | 4 |
| 1–5 | ~~IIII~~ I | 6 |

## Try These Together

1. Choose a scale for data from 3 to 32.
*HINT: Your scale must include 3 and 32.*

2. How many different whole number scores are possible in an interval from 25 to 30?
*HINT: Write each score, 25, 26, … and count how many, or subtract 30 − 25 and add 1.*

3. **Entertainment** Mr. Juarez awarded two points to each student answering the daily bonus question correctly. The data at the right lists the total number of points each student earned for the week. Make a frequency table for the data.

| | | | | | |
|---|---|---|---|---|---|
| 4 | 10 | 8 | 8 | 6 | 10 |
| 4 | 8 | 10 | 10 | 8 | 6 |
| 6 | 4 | 4 | 8 | 8 | 10 |
| 6 | 10 | 10 | 4 | 8 | 6 |

4. **Standardized Test Practice** What interval would you use in making a frequency table for this set of data?

2, 4, 3, 2, 10, 12, 8, 7, 5, 11

**A** 20 **B** 10 **C** 5 **D** 2

**Answers: 1.** Sample answer: 0–40 **2.** 6 **3.** See Answer Key. **4.** D

NAME ______________________ DATE ____________ PERIOD _____

# 2-2 Bar Graphs and Line Graphs (pages 56–59)

A graph represents data visually. A **bar graph** compares frequencies. A **line graph** compares changes over time.

| | |
|---|---|
| **Drawing a Vertical Bar Graph** | Draw and label the horizontal and vertical axes. Title your graph.<br>• Choose a scale and interval for the data and mark equal spaces on the vertical axis.<br>• Mark equal spaces on the horizontal axis and label the categories.<br>• Draw a bar for each category. The height shows the frequency. |
| **Drawing a Line Graph** | Draw and label the horizontal and vertical axes. Title your graph.<br>• Choose a scale and interval for the data and mark equal spaces on the vertical axis.<br>• Mark equal spaces on the horizontal axis and label the categories.<br>• Draw a dot to show the frequency for each category. Draw line segments to connect the dots. |

## EXAMPLES

**A** A class collects this data.

| Favorite Flavor | Frequency |
|---|---|
| vanilla | 13 |
| strawberry | 4 |
| chocolate | 10 |
| lemon | 2 |

Determine a scale for this data.

*The data go from 2 to 13. You might choose a scale from 0 to 15.*

**B** For the data in Example A, what would be a good interval?

*You could use an interval of 2 or 4.*

What are the labels for the categories on the horizontal axis?

*Vanilla, Strawberry, Chocolate, Lemon*

What is the label for the vertical axis? for the horizontal axis? for the graph?

*People; Flavors; Favorite Flavors*

### Try This Together

1. Make a bar graph for the data in Example A.
*HINT: You will have four bars. The tallest bar shows the most popular flavor.*

## PRACTICE

2. Make a line graph for the following set of data?

| Year | 1997 | 1998 | 1999 | 2000 |
|---|---|---|---|---|
| Number of Students in Drama Club | 3 | 9 | 17 | 15 |

3. **Standardized Test Practice** Estimate how many cars were sold in July.

**A** 15 **B** 35 **C** 25 **D** 10

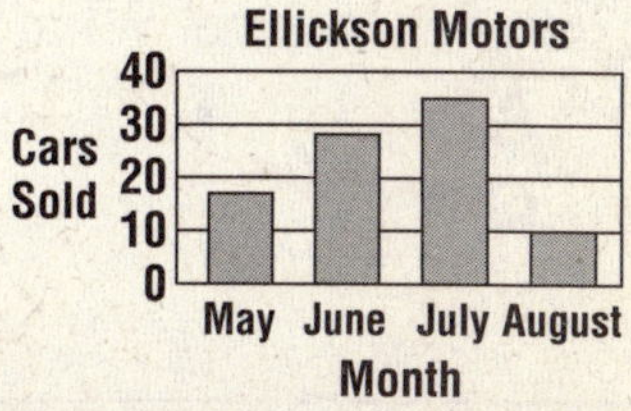


**Answers:** 1. See Answer Key. 2. See Answer Key. 3. B

NAME ______________________________ DATE ______________ PERIOD _____

# 2-3 Circle Graphs (pages 62–65)

A **circle graph** compares parts of a whole. The circle is the whole and the pie-shaped sections show the parts. All the percents in a circle graph add to 100%.

| **Reading a Circle Graph** | Read the title of the graph and the titles of all the sections.<br>• Recall that half of a circle is 50% and one-fourth is 25%.<br>• See how the percents match the sizes of the sections. |
|---|---|

## EXAMPLES

**A** The circle graph shows where the coins in Joel's collection come from. The percents are 10%, 20%, 30%, and 40%. Match each percent with the appropriate section of the graph.

*The section for Japan is the largest. It is almost one-half. So 40% of his coins come from Japan. The smallest section is Canada. So 10% of his coins come from Canada. The England section is larger than the Mexico one. So 30% come from England and 20% from Mexico.*

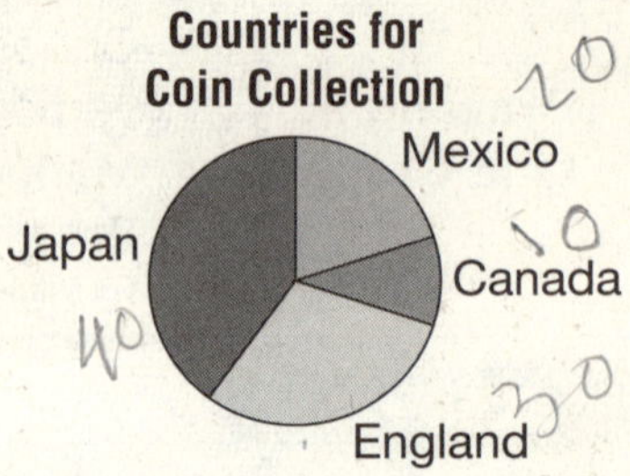


**B** What percent of his coins come from England and Mexico together?

*Add the percents: 30% added to 20% is 50%.*

### Try These Together

1. What fraction of Joel's collection comes from Canada and Japan together?
*HINT: What part of the circle are these two together?*

2. Canada and what other country together equal the same percent as Japan?
*HINT: Subtract the percent for Canada from that of Japan.*

## PRACTICE

***The circle graph shows the colors of homes in Anissa's neighborhood.***

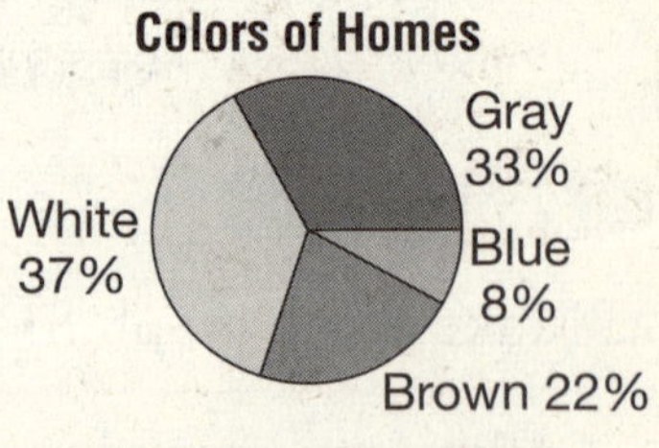


3. What percent of homes are blue?
4. What are the two most popular colors for homes in Anissa's neighborhood?

5. **Standardized Test Practice** The circle graph shows the pets students have. What percent of students do not have pets?

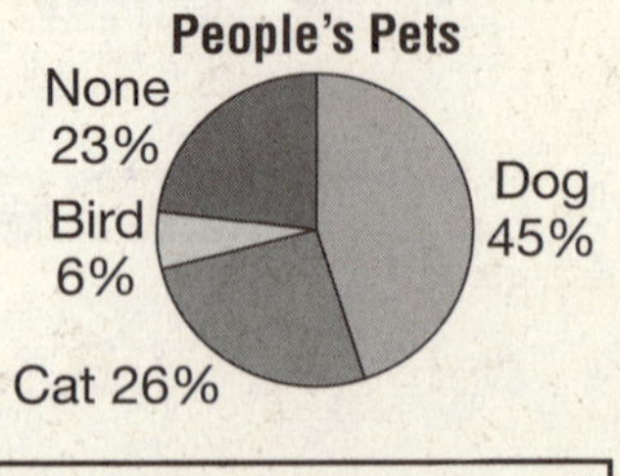


**A** 6% **B** 26%
**C** 23% **D** 45%

**Answers:** 1. $\frac{1}{2}$ 2. England 3. 8% 4. white and gray 5. C

NAME ______________________ DATE ____________ PERIOD _____

# 2-4 Making Predictions (pages 66–69)

You can use a line graph to help you make predictions.

| | |
|---|---|
| **Predicting with a Line Graph** | To make a prediction with a line graph,<br>• Extend the graph with a dashed line.<br>• From the point on the dashed line that shows where you want to make your prediction, draw a horizontal line to the left to meet the vertical axis.<br>• Read the value on the vertical axis. |

## EXAMPLES

The graph at the right shows how many books Kara and Bill read each month.

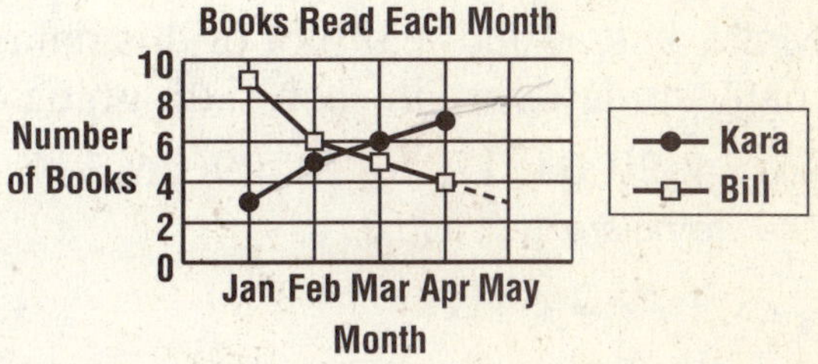


**A** What is the difference in April between the number of books Kara and Bill read?
*Kara read 7 and Bill read 4, so the difference is 3.*

**B** Predict how many books Bill will read in May.
*The extended line has a value on the vertical axis of 3 books.*

### Try These Together

1. Use the graph above to predict how many books Kara will read in May.
*HINT: Extend the line for Kara.*

2. How many more books would you expect Kara to read than Bill in May?
*HINT: Use your predictions for Kara and Bill.*

## PRACTICE

3. **Sports** The line graph shows how many laps Dominic swam each week for 6 weeks.

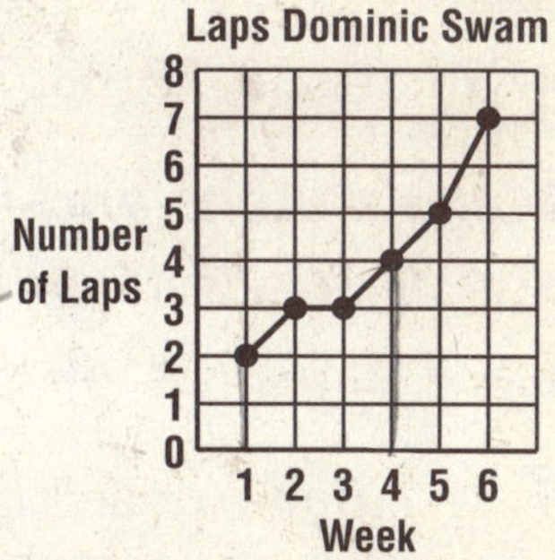


   **a.** Predict how many laps he will be able to swim in Week 7.
   **b.** How many more laps did he swim in Week 4 than in Week 1?
   **c.** Would you predict that Dominic will be able to swim more than 10 laps in Week 8?

4. **Standardized Test Practice** This line graph shows Jessica and Jared's math test scores for one week. Which day did they have the same score?

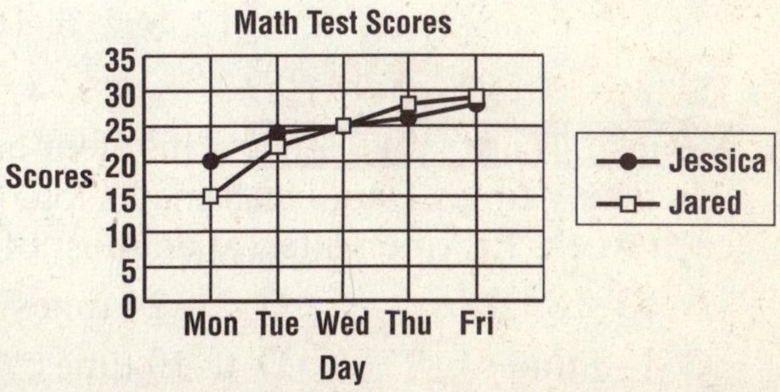


   **A** Monday  **B** Tuesday
   **C** Wednesday  **D** Friday

Answers: 1. 8 2. 5 3a. 9 3b. 2 3c. yes 4. C

NAME ________________ DATE ________ PERIOD ______

# 2-5 Stem-and-Leaf Plots (pages 72–75)

You can make a large data set easier to read with a **stem-and-leaf plot**. The **stems** are the tens digits. The **leaves** are the units digits.

| | |
|---|---|
| **Drawing a Stem-and-Leaf Plot** | Find the digits in the tens place for the least and the greatest numbers.<br>• Draw a vertical line and write the tens digits in order for the stems.<br>• Write the units digits, or leaves, to the right of their stems.<br>• Arrange the leaves in order from least to greatest. Include a key. |

## EXAMPLE

Make a stem-and-leaf plot of this data that shows how many students are in each sixth grade class.

15, 34, 20, 31, 17, 26, 24, 29, 26, 31

*The stems are 1, 2, and 3.*

| Stem | Leaf |
|---|---|
| 1 | 5 7 |
| 2 | 0 4 6 6 9 |
| 3 | 1 1 4 |

*1|5 = 15 students*

## Try These Together

1. How many classes are there in the data set in the Example?
*HINT: Count the numbers in the data set.*

2. What interval contains half of the class sizes?
*HINT: Which stem has the most leaves?*

## PRACTICE

***Determine the stems for each set of data.***

3. 13, 8, 12, 44, 26, 33, 15
4. 25, 64, 35, 22, 68, 71, 84, 14, 56, 41

***Make a stem-and-leaf plot for each set of data.***

5. 2, 5, 16, 22, 15, 14
6. 24, 25, 38, 34, 46, 58

7. **Aviation** Adrian's mother is an airline pilot. One week, he counted the number of hours she flew each day. Make a stem and leaf plot of the data.

12, 8, 2, 6, 10, 5

8. **Standardized Test Practice** This stem-and-leaf plot shows how many times Dara's classmates log on to the Internet each week. In which interval do most of the times fall?

**A** 12–18 times **B** 21–24 times
**C** 1–8 times **D** 0–10 times

| Stem | Leaf |
|---|---|
| 0 | 1 3 3 5 8 |
| 1 | 2 4 4 5 6 6 7 8 |
| 2 | 1 1 4 |

*2|1 = 21*

Answers: **1.** 10 **2.** 20–29 **3.** 0, 1, 2, 3, 4 **4.** 1, 2, 3, 4, 5, 6, 7, 8 **5–7.** See Answer Key. **8.** A

NAME ______________________ DATE ____________ PERIOD ______

# 2-6 Mean (pages 76–78)

One number used to represent an entire set of data is called a **measure of central tendency**. One of the most common measures of central tendency is the **mean**. The mean is also called the average.

| | |
|---|---|
| **Finding the Mean** | Add to find the sum of the data. Divide by the number of pieces of data. |

**EXAMPLE**

Find the mean of this set of data.

10, 13, 6, 7, 14, 28, 34, 5, 22, 11

*The sum of the data is 150. There are 10 pieces of data. Divide 150 by 10 to get a mean of 15.*

## Try This Together

1. The heights of students in Mr. Cohen's class are shown. Find the mean height.
*HINT: Find the sum, then divide.*

| Height (in.) | | | | |
|---|---|---|---|---|
| 58 | 55 | 50 | 64 | 53 |
| 62 | 66 | 54 | 57 | 62 |
| 60 | 55 | 59 | 65 | 64 |
| 56 | 53 | 62 | 57 | 68 |

**PRACTICE**

***Find the mean for each set of data.***

2. 10, 14, 18, 23, 10
3. 36, 24, 21, 58, 21
4. 22, 23, 29, 28, 24, 24
5. 11, 2, 4, 9, 4
6. 34, 46, 37
7. 9, 7, 3, 8, 2, 7

8. **Money Matters** Alicia is saving money for a portable CD player. The graph shows the costs of different CD players. What is the mean cost of the CD players?

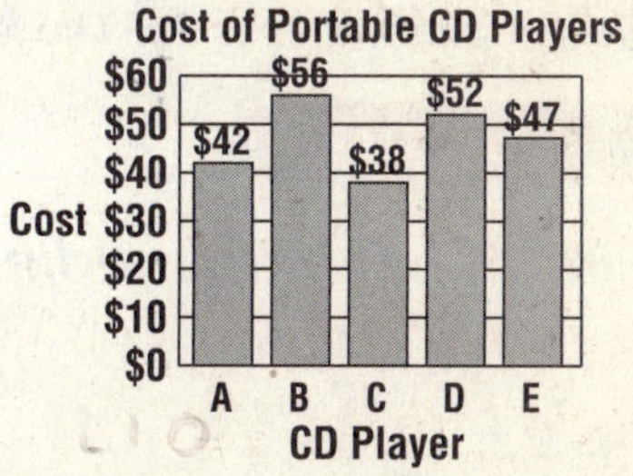


9. **Standardized Test Practice** What is the mean of the set of data in the table?

**A** 54 **B** 62
**C** 58 **D** 67

| School | Number of Students on Sports Teams |
|---|---|
| Blake | 56 |
| Irondale | 68 |
| River Trail | 101 |
| Jefferson | 43 |

Answers: 1. 59 in. 2. 15 3. 32 4. 25 5. 6 6. 39 7. 6 8. $47 9. D

NAME ______________________ DATE ____________ PERIOD ______

# 2-7 Median, Mode, and Range (pages 80–83)

You have already learned that the mean is one type of measure of central tendency. Other types are the **median**, the **mode**, and the **range**. The mean, median, and mode of a data set describe the center of a set of data. The range of a set of data describes how much the data vary.

| | |
|---|---|
| **Finding the Median** | Arrange the data in order from least to greatest. Find the middle number (or the mean of the two middle numbers). |
| **Finding the Mode** | Look for the number that appears most often. There may be more than one mode, or no mode. |
| **Finding the Range** | Subtract the least number in the data set from the greatest number in the data set. |

## EXAMPLE

The table shows the cost of 12 different DVDs.

| DVD Costs ($) | | | |
|---|---|---|---|
| 16 | 19 | 24 | 22 |
| 19 | 14 | 20 | 19 |
| 22 | 24 | 15 | 17 |

Find the median, mode, and range for the set of data.

To find the median, order the data from least to greatest.

14, 15, 16, 17, 19, 19, 19, 20, 22, 22, 24, 24

*Since there are two middle numbers, 19 and 19, find the mean of these numbers.*

*19 + 19 = 38, 38 ÷ 2 = 19 The median is 19.*

To find the mode, find the number or numbers that occur most often.

*The only number that occurs three times is 19. The mode is 19.*

To find the range, subtract the least value from the greatest value.

*The greatest value is 24. The least value is 14. So, the range is 24 − 14, or 10.*

## PRACTICE

***Find the mean, median, mode, and range for each set of data.***

**1.** 57, 51, 48, 63, 51 **2.** 86, 75, 88 **3.** 9, 18, 9, 17, 9, 10

**4.** 22, 19, 31, 28 **5.** 36, 35, 42, 35, 42 **6.** 2, 11, 6, 1

**7.** 66, 59, 75, 72, 65, 59 **8.** 2, 9, 1, 1, 2 **9.** 97, 54, 89

**10.** **Standardized Test Practice** Which measure of central tendency may not apply to a set of data?

**A** mean **B** median **C** mode **D** range

**Answers: 1.** 54; 51; 51; 15 **2.** 83; 86; no mode; 13 **3.** 12; 9.5; 9; 9 **4.** 25; 25; no mode; 12 **5.** 38; 36; 35 and 42; 7 **6.** 5; 4; no mode; 10 **7.** 66; 65.5; 59; 16 **8.** 3; 2; 1 and 2; 8 **9.** 80; 89; no mode; 43 **10.** C

NAME ______________________ DATE ____________ PERIOD ______

# 2-8 Analyzing Graphs (pages 86–89)

Graphs are sometimes drawn to influence conclusions by misrepresenting the data.

| Determining when a Graph is Misleading | • Is there is a label on both scales and a title on the graph?<br>• Does the scale start at zero?<br>• The mean is best to represent data that are grouped closely together.<br>• The median is best for widely scattered data.<br>• The mode is best for data that have several repeated data values. |
|---|---|

## EXAMPLES

**A** What measure of central tendency would best represent the ages of people in your math class? *Many of the ages will be repeated. The mode is best.* Mode

**B** What measure would best represent the annual salaries in a large company? *The salaries are widely scattered. Choose the median.* Median

### Try These Together

1. What measure best represents the distance each student lives from school? *HINT: Are the data values fairly close together?*

2. Is the mode for a set of data always one of the data values? *HINT: Remember the definition of mode.*

Mean  yes

## PRACTICE

**Fitness** ***The graphs display the same data for prices at the Fitness Center.***

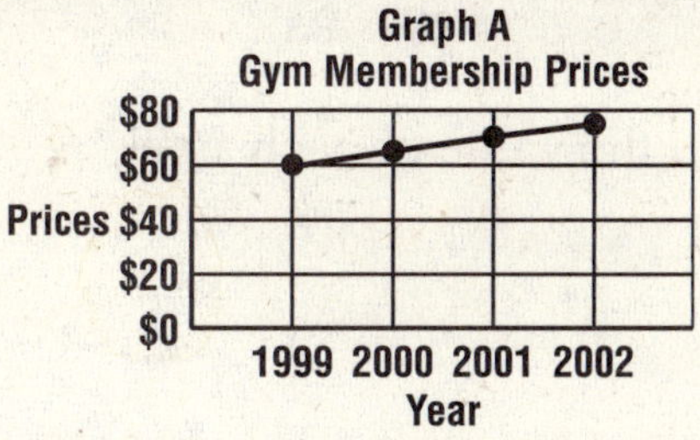


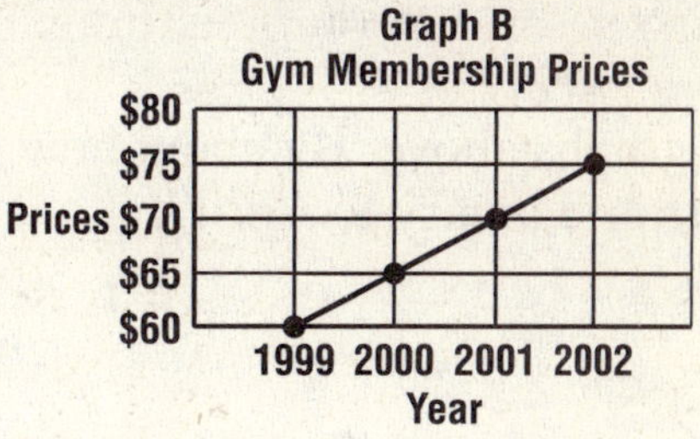


3. If someone were trying to sell memberships by saying that it will cost a lot more in the future, which graph might be used? Graph B

4. Why is graph B misleading?

5. **Standardized Test Practice** The results of a class survey on the number of hours each student spends on homework every night are shown in the table. What is the mode for this set of data?

| Number of Hours | Frequency |
|---|---|
| 1 | 4 |
| 2 | 8 |
| 3 | 2 |
| 4 | 3 |

**A** 1 **B** 2 **C** 4 **D** 8

**Answers:** 1. mean 2. yes 3. Graph B 4. It does not show $0 with a break in the vertical axis between $0 and $60. 5. B

# 2 Chapter 2 Review

## Stock Market Game

In a stock market game, teams of students must pick a stock to "buy." After several months, the team whose stock gains the most value wins. Teams make their decisions about which stocks to buy based on the price of the stock over the past several months. Use the information below to help your team pick the best stock.

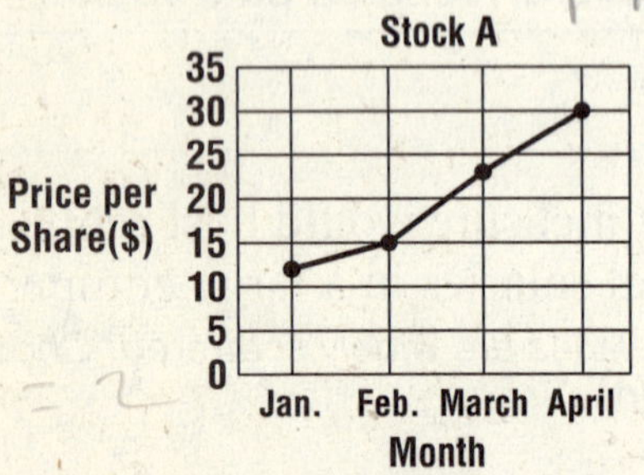


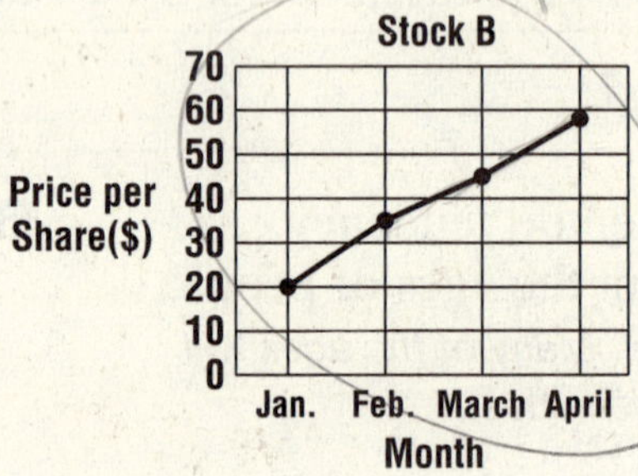


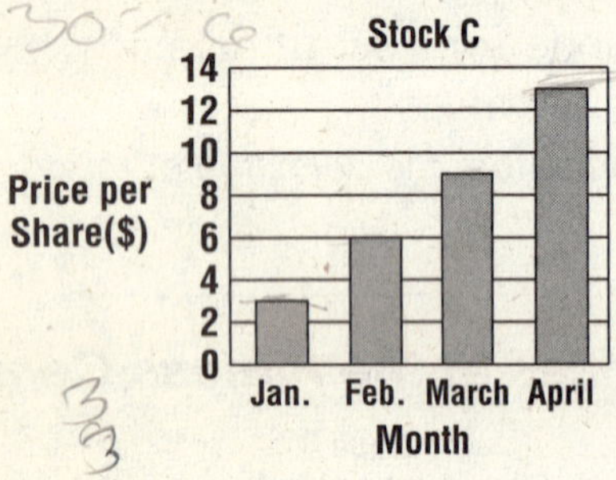


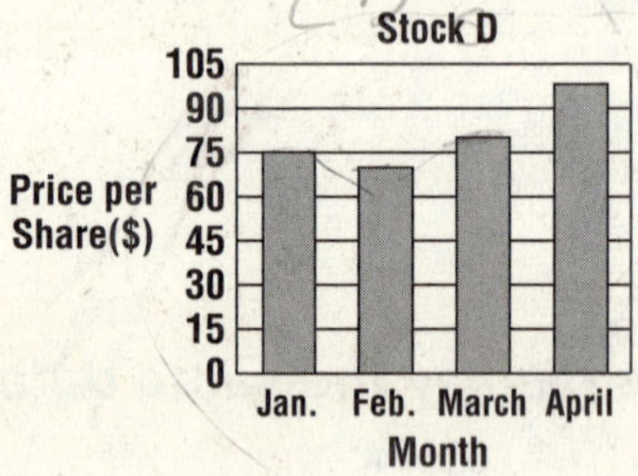


1. Read the graphs above. By about how much did the value of each stock increase from January to April?

2. To win the stock market game, you want to buy the stock that will increase in value the most over the next several months. Based on the amount that each stock has increased in value, which stock would you want your team to buy? Explain.

Answers are located on p. 105.

NAME ______________________ DATE __________ PERIOD ____

# 3-1 Representing Decimals (pages 102–105)

Decimals are numbers that are expressed using a decimal point. The decimal point separates the whole number part of the decimal from the part that is less than one. You use **place-value** positions to name decimals.

Decimals can be written in standard form and expanded form. **Standard form** is the usual way to write a number. **Expanded form** is a sum of the products of each digit and its place value.

| | |
|---|---|
| **word form** | eighteen hundredths |
| **standard form** | 0.18 |
| **expanded form** | $(1 \times 0.1) + (8 \times 0.01)$ |

## EXAMPLES

**A** Use the place-value chart at the right to help you write twenty-three thousandths as a decimal.

*Write the digits 2 and 3 so that the 3 is in the thousandths place. Fill in zeroes to the left through the ones place: twenty-three thousandths is written as 0.023.*

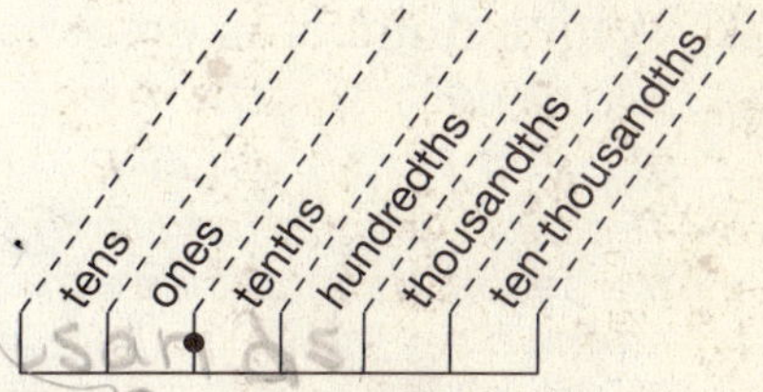


**B** Write 0.0012 in word form.

*The 2 is in the ten-thousandths place. 0.0012 is twelve ten-thousandths.*

## Try These Together

1. Write thirty and three hundredths as a decimal.
   *HINT: The word "and" tells you the location of the decimal point.*
2. Write 52 and 4 thousandths as a decimal.
   *HINT: Write the whole number part (52) starting in the tens place. Use zeros to fill in the tenths and hundredths places.*

## PRACTICE

***Write each decimal in word form.***

3. 0.5
4. 0.08
5. 0.007
6. 1.2
7. 5.02
8. 2.3
9. 17.1
10. 0.65

***Write each decimal in standard form and in expanded form.***

11. five hundredths
12. eighty-five thousandths
13. two tenths
14. **Health** A human's normal body temperature is ninety-eight and six tenths degrees. Write ninety-eight and six tenths as a decimal.
15. **Standardized Test Practice** Which decimal represents eight and nine hundredths?

**A** 0.89 **B** 8.9 **C** 8.09 **D** 89.9

**Answers: 1.** 30.03 **2.** 52.004 **3.** five tenths **4.** eight hundredths **5.** seven thousandths **6.** one and two tenths **7.** five and two hundredths **8.** two and three tenths **9.** seventeen and one tenth **10.** sixty-five hundredths **11.** 0.05; $(0 \times 0.1) + (5 \times 0.01)$ **12.** 0.085; $(0 \times 0.1) + (8 \times 0.01) + (5 \times 0.001)$ **13.** 0.02; $(2 \times 0.1)$ **14.** 98.6 **15.** C

# 3-2 Comparing and Ordering Decimals (pages 108–110)

You can compare decimals by comparing the digits in each place-value position or by placing the decimals on a number line. Recall that < means *less than* and > means *greater than.*

| | |
|---|---|
| **Comparing Decimals** | Line up the decimal points of the two numbers you want to compare. Then starting at the left, compare the digits in the same place-value position. When you come to a place where the digits are not equal, the decimal with the greater digit is the greater decimal number. On a number line, numbers to the right are greater than numbers to the left. |

## EXAMPLES

**A** Which number is greater, 1.09 or 1.9?

*1.09*
*1.9*

*The digits are the same in the ones place but the second number has a greater digit in the tenths place, so 1.9 is the greater number.*
*1.9 > 1.09*

**B** Order 21.98, 24.03, 2.4, and 2.198 from least to greatest.

*21.98*
*24.03*
*2.4*
*2.198*

*2.198, 2.4, 21.98, 24.03*

## Try These Together

1. Which of these numbers is to the left of 4.5 on a number line: 40.5 or 4.05?
*HINT: Which number is less than 4.5?*

2. Order 0.01, 0.002, and 0.02 from greatest to least.
*HINT: You can also look at hundredths as money. Which is greater, 2 cents or 1 cent?*

## PRACTICE

***Use >, <, or = to compare each pair of decimals.***

3. 0.41 ● 0.45
4. 1.8 ● 1.80
5. 8.25 ● 8.31
6. 46.85 ● 46.96
7. 0.06 ● 0.61
8. 0.78 ● 0.45
9. 1.363 ● 1.367
10. 458.6 ● 458.4
11. 1.03 ● 1.01

***Order each set of decimals from least to greatest.***

12. 12.56, 12.58, 12.36, 12.41
13. 456.9, 455.8, 455.4, 456.3
14. Which is the greatest, 5.06, 5.60, or 5.006?
15. **Standardized Test Practice** Which of these numbers is the smallest: 4.015, 4.014, 4.018, or 4.011?

**A** 4.011 **B** 4.014 **C** 4.018 **D** 4.015

Answers: 1. 4.05 2. 0.02, 0.01, 0.002 3. < 4. = 5. < 6. < 7. < 8. > 9. < 10. > 11. > 12. 12.36, 12.41, 12.56, 12.58 13. 455.4, 455.8, 456.3, 456.9 14. 5.60 15. A

NAME ______________________________ DATE ____________ PERIOD _____

# 3-3 Rounding Decimals (pages 111–113)

You can round decimals to any place-value position.

| | |
|---|---|
| **Rounding Decimals** | • Underline the digit to be rounded.<br>• Look at the digit to the right of the place being rounded.<br>• Leave the underlined digit the same if the digit to the right is 0, 1, 2, 3, or 4.<br>• Round up by adding 1 to the underlined digit if the digit to the right is 5, 6, 7, 8, or 9.<br>• Then drop all the digits to the right of the underlined digit. |

## EXAMPLES

**A** Round 25.0743 to the nearest tenth.

*Underline the digit in the tenths place (0). Look at the digit to the right (7). Since 7 is greater than 5, add one to the 0. Then drop all the digits to the right. 25.1*

**B** Round 324.67 to the nearest ten.

*Underline the digit in the tens place (2). Because the next digit to the right is less than 5, leave the 2 the same. Replace the 4 with a 0 to keep the digits to the left of the decimal in the proper places. Drop the digits to the right of the decimal. 320*

### Try These Together

**1.** Round $6.50 to the nearest dollar.
*HINT: Remember that with a 5 you round up.*

**2.** Is 0.345 closer to 0.3 or 0.4?
*HINT: Use zeros to write each number with the same number of decimal places.*

## PRACTICE

***Round each decimal to the indicated place-value position.***

**3.** 1.21; tenths
**4.** 8.63; ones
**5.** 38.622; hundredths
**6.** 4.37; tenths
**7.** 24.8568; thousandths
**8.** 27.53; ones
**9.** 13.58; tenths
**10.** 23.2594; thousandths
**11.** 99.3482; thousandths
**12.** 95.524; hundredths
**13.** 9.64; tenths
**14.** 87.635; hundredths

**15.** Round 67.687 to the nearest tenth.

**16.** Round $12.35 to the nearest dollar.

**17. Entertainment** It costs $3.99 to rent a movie from the video store. If you rented a movie, how much would you probably say it cost? (Round $3.99 to the nearest dollar.)

**18. Standardized Test Practice** People in the United States are living longer than ever before. The average life span is 76.1 years. What is this number rounded to the nearest year?

**A** 77 **B** 76.2 **C** 76.1 **D** 76

**Answers:** **1.** $7 **2.** 0.3 **3.** 1.2 **4.** 9 **5.** 38.62 **6.** 4.4 **7.** 24.857 **8.** 28 **9.** 13.6 **10.** 23.259 **11.** 99.348 **12.** 95.52 **13.** 9.6 **14.** 87.64 **15.** 67.7 **16.** $12.00 **17.** $4 **18.** D

NAME ______________________ DATE ____________ PERIOD _____

# 3-4 Estimating Sums and Differences

(pages 116–119)

**Rounding**, **front-end estimation**, and **clustering** are all ways to estimate.

| | |
|---|---|
| **Estimating by Rounding** | • Round each number to the same place-value position, often ones.<br>• Add or subtract the rounded numbers. |
| **Front-End Estimation** | • Add or subtract the front digits.<br>• Add or subtract the digits in the next place value position. |
| **Estimating by Clustering** | Use clustering when all the numbers are close to the same number.<br>• Round each number to the same number—the number they cluster around.<br>• Add or subtract the rounded numbers. |

## EXAMPLES

**A** Estimate using rounding.
\$45.27 − \$4.87
*Round each amount to the nearest dollar.*
*\$45 − \$5 = \$40*

**B** Estimate using clustering.
10.76 + 11.1 + 10.98 + 11 + 10.7
*All the numbers cluster around 11, so add*
*11 + 11 + 11 + 11 + 11 = 55.*

### Try These Together

1. About how much more is \$25.10 than \$14.98?
*HINT: Round each amount to the nearest dollar and subtract.*

2. About how much lower is a temperature of 59.5 degrees than one of 91.3 degrees?
*HINT: Round before you subtract.*

## PRACTICE

***Estimate using rounding.***

3. 0.76 + 0.14
4. 5.3 + 4.8
5. 25.6 − 3.8

***Estimate using front-end estimation.***

6. 26.4 + 13.5
7. 57.35 − 34.68
8. 18.25 + 31.95

***Estimate using clustering.***

9. \$6.12 + \$5.87
10. 0.86 + 0.9 + 0.93
11. 2.9 + 3.2 + 3.1

12. **Money Matters** Keesha is going out for pizza with her friends. She knows pizza will cost \$5.65 and a drink will cost \$1.55. Estimate how much money she should bring with her.

13. **Standardized Test Practice** Thomas needs 1.2 pounds of chocolate chips and 0.8 pounds of peanut butter chips. Estimate how many pounds of chocolate and peanut butter chips he needs all together.

**A** 1 **B** 2 **C** 3 **D** 4

**Answers: 1.** about \$10 **2.** about 30 degrees **3.** 0.9 or 1 **4.** 10 **5.** 22 **6.** 39.0 **7.** 23.00 **8.** 49.00 **9.** \$12.00 **10.** 3 **11.** 9 **12.** \$8.00 **13.** B

NAME ______________________ DATE __________ PERIOD _____

# 3-5 Adding and Subtracting Decimals

(pages 121–124)

You add and subtract decimals the same way you do whole numbers, after you line up the decimal points.

| | |
|---|---|
| **Adding and Subtracting Decimals** | • Write the numbers you want to add or subtract so that the decimal points are in a line. Add zeros if they are needed.<br>• Estimate the sum or difference so you can check to see if your final answer is reasonable.<br>• Add or subtract. Compare the result with your estimate. |

## EXAMPLES

**A** Find the sum of 2.45 and 30.7.

*Line up the decimal points and add a zero.*

$$\begin{array}{r} 2.45 \\ +\ 30.70 \\ \hline 33.15 \end{array}$$

*Estimate first.*
*This is about 31 + 2 or 33.*
*This is reasonably close to the estimate of 33.*

**B** Subtract 27.8 from 60.

*Line up the decimal points and add a zero.*

$$\begin{array}{r} 60.0 \\ -\ 27.8 \\ \hline 32.2 \end{array}$$

*Estimate first.*
*This is about 60 − 30 or 30.*
*This is reasonably close to the estimate of 30.*

## Try These Together

**1.** Subtract 3 − 2.09.
*HINT: Remember that 3 is the same as 3.00*

**2.** Add 4.56 + 23.
*HINT: Rewrite 23 with a decimal point and two zeros as you line up the numbers to add.*

## PRACTICE

***Add or subtract.***

**3.** 5.6 + 4.2
**4.** 1.25 + 1.34
**5.** 12.61 + 3.27
**6.** 25.69 − 24.54
**7.** 2.7 − 1.1
**8.** 13.32 − 9.12
**9.** \$10.26 − \$8.28
**10.** 5.68 + 3.45
**11.** 9 + 3.43
**12.** 4.05 + 2.68
**13.** 16.51 − 13.25
**14.** 0.06 + 0.15

**15.** What is the value of $c + d$ if $c = 22.4$ and $d = 36.2$?

**16.** Evaluate $q - r$ if $q = 3.5$ and $r = 2.1$.

**17.** **Surveys** Manuel surveyed two of his friends to find out the average number of sodas they drink in one week. Carl drinks 4.5 sodas and Jon drinks 6.75 sodas. How many sodas do Carl and Jon drink together in one week?

**18.** **Standardized Test Practice** Janette is 1.55 meters tall and Kirsten is 1.47 meters tall. How much taller is Janette than Kirsten?

**A** 0.08 m **B** 0.06 m **C** 0.07 m **D** 0.09 m

Answers: **1.** 0.91 **2.** 27.56 **3.** 9.8 **4.** 2.59 **5.** 15.88 **6.** 1.15 **7.** 1.6 **8.** 4.2 **9.** \$1.98 **10.** 9.13 **11.** 12.43 **12.** 6.73 **13.** 3.26 **14.** 0.21 **15.** 58.6 **16.** 1.4 **17.** 11.25 **18.** A

NAME ______________ DATE ______________ PERIOD ______

# 3 Chapter 3 Review

## Decimal Derby

This year's mule derby had 8 mules running a quarter-mile race. The finishing times are given below.

| Mule | 1 | 2 | 3 | 4 | 5 | 6 | 7 | 8 |
|---|---|---|---|---|---|---|---|---|
| Time (sec) | 52.206 | 58.671 | 51.992 | 52.187 | 52.037 | 52.945 | 55.473 | 53.628 |

1. Place the mules in the order in which they finished the race.

   51.992 52.037 52.187 52.206 52.945 53.628 55.473 58.671

2. What was the time difference between the first and second place mules?

   00.045

3. What was the time difference between the second and third place mules?

   00.150

4. How many seconds were there between the time the first place mule finished, and the time the last place mule finished?

   06.679

5. What were the finishing times of the first three mules, rounded to the nearest tenth?

   52.000 52.000 52.200

6. The mules' names are in the table below. Use the mules' names and the order in which they finished the race to complete the sentences below.

| Mule | 1 | 2 | 3 | 4 | 5 | 6 | 7 | 8 |
|---|---|---|---|---|---|---|---|---|
| Name | If You Just Try | You and Me | Working | Fun and Easy | Decimals | Little | Easy Math | Hard Work |

Working with Decimals is Fun and Easy

If you just try a little. hard work makes

easy math for You an d me.

Answers are located on p. 105.

NAME ______________________ DATE ____________ PERIOD _____

# 4-1 Multiplying Decimals by Whole Numbers (pages 135–138)

When you multiply a decimal by a whole number, you can estimate to find where to put the decimal point in the product. You can also place the decimal point by counting the decimal places in the decimal factor.

| | |
|---|---|
| **Estimation** | • Estimate the product of a decimal and a whole number by rounding the decimal to its greatest place value position and then multiplying.<br>• Multiply as you do with whole numbers.<br>• Use your estimate as a guide for placing the decimal in the product. |
| **Counting Decimal Places** | • Multiply the decimal and whole number as if they were both whole numbers.<br>• Count the number of decimal places in the decimal factor. Place the decimal point in the answer so that there are the same number of decimal places as in the decimal factor. Annex (or write) zeros to the left of your answer if more decimal places are needed. |

## EXAMPLES

***Find the value of each expression.***

**A** Find 22.3 × 5.

20 × 5 *Round the decimal. Estimate the product; 100.*

22.3
× 5 *Multiply as with whole numbers.*
111.5 *Use the estimate, 100, as a guide to placing the decimal. Place the decimal point after 111.*

**B** Find 0.015 × 3.

0.015 *There are 3 decimal places in this*
× 3 *factor.*
0.045 *Annex a zero on the left to make three decimal places.*

### Try These Together

***Multiply.***

1. 4.02 × 5

*HINT: Estimate the product; then, multiply as with whole numbers.*

2. 0.017 × 2

*HINT: Count the decimal places in the decimal factor.*

## PRACTICE

***Multiply.***

3. 0.4 × 9
4. 0.62 × 7
5. 1.71 × 3
6. 3.65 × 5
7. 61 × 0.004
8. 9.7 × 561
9. 5,618 × 6.83

10. **Standardized Test Practice** Evaluate 104*h* if *h* = 7.1.

**A** 0.7384 **B** 738.4 **C** 7,384 **D** 73,840

**Answers:** **1.** 20.1 **2.** 0.034 **3.** 3.6 **4.** 4.34 **5.** 5.13 **6.** 18.25 **7.** 0.244 **8.** 5,441.7 **9.** 38,370.94 **10.** B

# 4-2 Multiplying Decimals (pages 141–143)

When you multiply two decimals, multiply as with whole numbers. To place the decimal point, find the sum of the number of decimal places in each factor. The product has the same number of decimal places.

## EXAMPLES

***Find the value of each expression.***

**A** Find 2.9 × 4.1.

3 × 4 *Round the decimals. Estimate the product; 12.*

```
   2.9   one decimal place
 × 4.1   one decimal place
    29
  11 6
 11.89   two decimal places
```

*The product is 11.89. Compared to the estimate, the product is reasonable.*

**B** Find 3.2 × 5.7.

3 × 6 *Round the decimals. Estimate the product; 18.*

```
   3.2   one decimal place
 × 5.7   one decimal place
   224
  160
 18.24   two decimal places
```

*The product is 18.24. Compared to the estimate, the product is reasonable.*

## Try These Together

***Multiply.***

**1.** 7.6 × 2.3

*HINT: Estimate the product. Then multiply as with whole numbers.*

**2.** 0.52 × 2.6

*HINT: Count the decimal places in the factors.*

## PRACTICE

***Multiply.***

**3.** 0.52 × 1.7 **4.** 6.6 × 0.054 **5.** 2.73 × 5.86

**6.** 1.5 × 6.4 **7.** 0.9 × 0.036 **8.** 3.25 × 7.3

**9.** 0.85 × 0.04 **10.** 4.6 × 8.2 **11.** 12.6 × 2.7

**12.** Find $2.5a + b$ if $a = 4.65$ and $b = 5.8$

**13. Standardized Test Practice** Multiply 1.6 × 0.023.

**A** 0.0368 **B** 0.368 **C** 3.68 **D** 36.8

**Answers: 1.** 17.48 **2.** 1.352 **3.** 0.884 **4.** 0.3564 **5.** 15.9978 **6.** 9.6 **7.** 0.0324 **8.** 23.725 **9.** 0.034 **10.** 37.72 **11.** 34.02 **12.** 17.425 **13.** A

NAME ______________________ DATE __________ PERIOD _____

# 4-3 Dividing Decimals by Whole Numbers (pages 144–147)

When you divide a decimal by a whole number, place the decimal point in the quotient directly above the decimal point in the dividend. Then, divide as you do with whole numbers.

**EXAMPLES**

***Find each quotient.***

**A** 14.8 ÷ 2

*First estimate: 14 ÷ 2 = 7.*
*Place the decimal point.*

$$\begin{array}{r} 7.4 \\ 2\overline{)14.8} \\ \underline{-14} \\ 8 \\ \underline{-8} \\ 0 \end{array}$$

*Divide as with whole numbers.*

**B** 27.3 ÷ 3

*First estimate: 27 ÷ 3 = 9.*
*Place the decimal point.*

$$\begin{array}{r} 9.1 \\ 3\overline{)27.3} \\ \underline{-27} \\ 3 \\ \underline{-3} \\ 0 \end{array}$$

*Divide as with whole numbers.*

## Try These Together

***Find each quotient.***

**1.** 25.4 ÷ 2
*HINT: Use the dividend as a guide to placing the decimal in the quotient.*

**2.** 6.16 ÷ 4
*HINT: Use the dividend as a guide to placing the decimal in the quotient.*

**PRACTICE**

***Divide. Round to the nearest tenth if necessary.***

**3.** $7\overline{)29.4}$ **4.** $12\overline{)915.96}$ **5.** $31\overline{)570.4}$

**6.** 155.1 ÷ 66 **7.** $17\overline{)152.83}$ **8.** $42\overline{)68.46}$

**9.** 81.81 ÷ 27 **10.** 41.79 ÷ 86 **11.** $21\overline{)698.44}$

**12.** $69\overline{)73.67}$ **13.** 58.42 ÷ 16 **14.** 247.73 ÷ 44

**15.** 104.745 ÷ 34 **16.** $65\overline{)623.86}$ **17.** $91\overline{)5.237}$

**18.** 24.15 ÷ 7 **19.** 1.507 ÷ 11 **20.** 144.96 ÷ 48

**21. Money Matters** Mika borrowed $18.30 from his parents to buy a book. How much should Mika give his parents each week if he plans to make equal payments for six weeks?

**22. Standardized Test Practice** Round 126.33 ÷ 16 to the nearest hundredth.

**A** 7.8 **B** 7.89 **C** 7.90 **D** 7.93

Answers: 1. 12.7 2. 1.54 3. 4.2 4. 76.3 5. 18.4 6. 2.4 7. 9.0 8. 1.6 9. 3.0 10. 0.5 11. 33.3 12. 1.1 13. 3.7 14. 5.6 15. 3.1 16. 9.6 17. 0.1 18. 3.5 19. 0.1 20. 3.0 21. $3.05 22. C

NAME ______________________ DATE ____________ PERIOD ______

# 4-4 Dividing by Decimals (pages 152–155)

When you divide decimals by decimals, you must change the divisor to a whole number. To do this, multiply both the divisor and dividend by the same power of 10. Then divide as with whole numbers.

## EXAMPLES

***Find each quotient.***

**A** 4.4 ÷ 2.5

*First estimate: 4 ÷ 2 = 2*

```
              1.76
2.5)4.4    25)44.00
             -25
              190
             -175
              150
             -150
                0
```

*Multiply the dividend and divisor by 10. Place the decimal point.*

*Divide as with whole numbers.*

**B** Find 33.08 ÷ 16.2 to the nearest hundredth.

```
                    2.041
16.2)33.08    162)330.800
                 -324
                    68
                    -0
                   680
                  -648
                   320
                  -162
                   158
```

*Divide to the thousandths place to round to the nearest hundredth. Since 68 is less than the divisor, write a zero in the quotient. To the nearest hundredth, the quotient is 2.04.*

### Try These Together

***Divide.***

1. 5.4 ÷ 1.2
   *HINT: Multiply the dividend and divisor by the same power of 10.*

2. 16.646 ÷ 4.1
   *HINT: Do not forget to fill in spaces in the quotient with zeros.*

## PRACTICE

***Divide.***

3. 3.9)849.03
4. 5.97)3,826.77
5. 11.5)634.11
6. 0.15 ÷ 0.008
7. 6.8034 ÷ 6.67
8. 8.814 ÷ 0.0678

***Find each quotient to the nearest hundredth.***

9. 0.31)9.4
10. 17.6)21.191
11. 8.39)486.7
12. 63.66 ÷ 7.23
13. 1.76 ÷ 28
14. 59.681 ÷ 0.98

15. **Hobbies** Paquita wants to make a necklace 55.9 cm long using beads with a diameter of 1.3 cm. How many beads does she need?

16. **Standardized Test Practice** Find 4.998 ÷ 3.4.

**A** 1.47 **B** 1.52 **C** 6.82 **D** 16.99

Answers: **1.** 4.5 **2.** 4.06 **3.** 217.7 **4.** 641 **5.** 55.14 **6.** 18.75 **7.** 1.02 **8.** 130 **9.** 30.32 **10.** 1.20 **11.** 58.01 **12.** 8.80 **13.** 0.06 **14.** 60.90 **15.** 43 beads **16.** A

NAME ______________________ DATE ____________ PERIOD _____

# 4-5 Perimeter (pages 158–160)

The **perimeter** ($P$) of a closed figure is the distance around the figure. You can find the perimeter by adding the measures of all the sides of the figure.

| | | |
|---|---|---|
| **Perimeter of a Rectangle** | The perimeter of a rectangle is two times the length $\ell$ plus two times the width $w$, or $P = 2\ell + 2w$. | 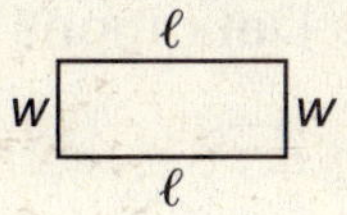 |
| **Perimeter of a Square** | The perimeter of a square is four times the measure of any of its sides $s$, or $P = 4s$. | 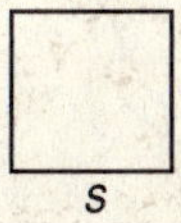 |



## EXAMPLES

**A** Find the perimeter of a rectangle with a length of 12.3 ft and a width of 6 ft.

$P = 2\ell + 2w$
$P = 2(12.3) + 2(6)$ — *$\ell = 12.3$ and $w = 6$*
$P = 24.6 + 12$
$P = 36.6$ — *The perimeter is 36.6 ft.*

**B** Find the perimeter of a square whose sides measure 3 yd.

$P = 4s$
$P = 4(3)$ — *$s = 3$*
$P = 12$ — *The perimeter is 12 yd.*

## Try These Together

1. Find the perimeter of a rectangle with a length of 9 m and a width of 4 m.
   *HINT: The perimeter is two times the length plus two times the width.*
2. Find the perimeter of a square whose sides measure 8 in.
   *HINT: Perimeter of a square is four times any side.*

## PRACTICE

***Find the perimeter of each figure.***

3. 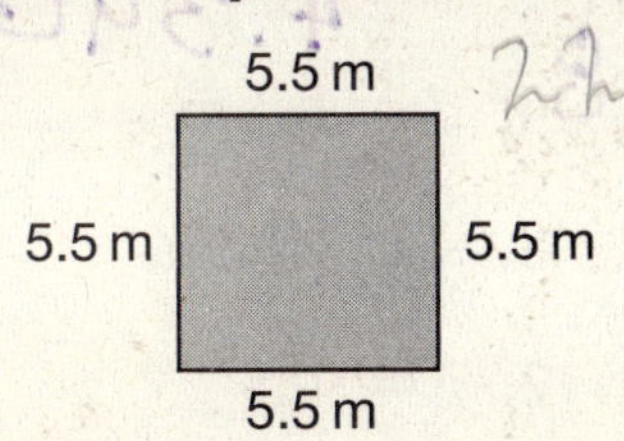


4. 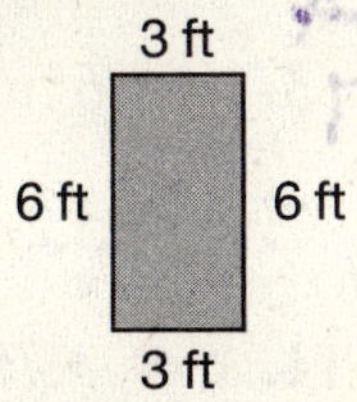


5. 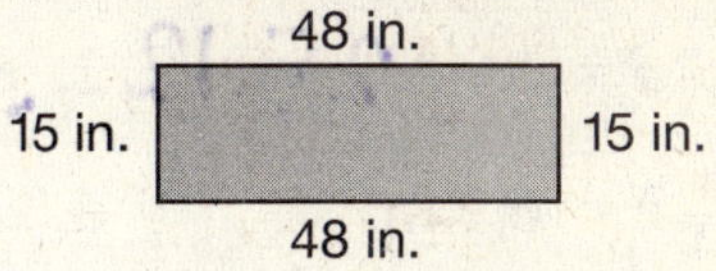


6. square: $s = 18.4$ cm
7. rectangle: $\ell = 12$ yd; $w = 8$ yd
8. square: $s = 11.6$ ft

9. **Standardized Test Practice** A rectangle is 8.6 cm long, and its perimeter is 18 cm. What is its width?
   **A** 9.4 cm  **B** 2.09 cm  **C** 0.8 cm  **D** 0.4 cm

Answers: 1. 26 m 2. 32 in. 3. 22 m 4. 18 ft 5. 126 in. 6. 73.6 cm 7. 40 yd 8. 46.4 ft 9. D

NAME ______________________________ DATE ____________ PERIOD _____

# 4-6 Circumference (pages 161–164)

A **circle** is a set of points in a plane, all of which are the same distance from a fixed point in the plane called the **center**.

| | |
|---|---|
| **Circle Definitions** 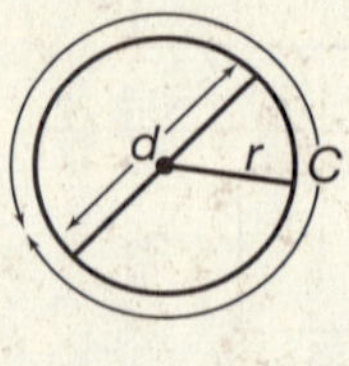 | • The distance from the center of a circle to any point on the circle is called the **radius *r***.<br>• The distance across the circle through the center is called the **diameter *d***. The diameter of a circle is twice the length of its radius.<br>• The **circumference *C*** is the distance around the circle.<br>• The circumference of a circle is always a little more than three times its diameter. The exact number of times is represented by the Greek letter $\pi$ (pi). The decimal 3.14 and the fraction $\frac{22}{7}$ are used as approximations for $\pi$. |
| **Finding the Circumference** | The circumference of a circle is equal to $\pi$ times the diameter or $\pi$ times twice its radius, $C = \pi d$ or $C = 2\pi r$. |

## EXAMPLE

Find the circumference of a circle with a diameter of 2.5 in.

$C = \pi d$

$\approx 3.14 \cdot 2.5$ *Replace $\pi$ with 3.14 and d with 2.5.*

$\approx 7.85$ *Multiply.*

*The circumference of the circle is about 7.85 inches.*

## PRACTICE

***Find the circumference of each circle described. Round to the nearest tenth.***

**1.** $d = 8$ in. 25.12 **2.** $r = 4.25$ ft ~~26.64~~ 26.7 **3.** $r = 6$ m 37.68 **4.** $d = 1.4$ m 4.396

**5.** $r = 0.9$ in. 5.7 **6.** $d = 2.5$ ft 7.9 **7.** $r = 5.2$ in. 32.7 **8.** $d = 10$ cm 31.4

**9.** $d = 7.5$ m 23.6 **10.** $r = 22$ cm 138.2 **11.** $d = 3.75$ yd 11.8 **12.** $r = 9$ ft 56.5

**13.** **Standardized Test Practice** The Sacagawea Golden Dollar coin has a radius of 13.25 mm. What is its circumference?

**A** 41.2 mm **B** 83.3 mm **C** 26.5 mm **D** 79.5 mm

**Answers: 1.** 25.12 in. **2.** 26.7 ft **3.** 37.68 m **4.** 4.396 m **5.** 5.7 in. **6.** 7.9 ft **7.** 32.7 in. **8.** 31.4 cm **9.** 23.6 m **10.** 138.2 cm **11.** 11.8 yd **12.** 56.5 ft **13.** B

NAME ______________________ DATE ____________ PERIOD _____

# 4 Chapter 4 Review

## Decimal Treasure Hunt

Every week, Mr. Jefferson records extra credit for the first person in his math class who can locate the hidden treasure in his room. The hidden treasure is on a bulletin board on the back of a card with a certain number on it. There are many cards on the bulletin board, so the students first solve a set of problems in order to find the hidden treasure and earn the extra credit.

The following problems will help you find this week's treasure.

1. Start with the number 12.32. Multiply this number by 4.

2. Take your answer from problem 1 and add it to 3(4 + 6).

3. Multiply the answer from problem 2 by 2.3.

4. Divide the answer from problem 3 by 8.

5. Divide the answer from problem 4 by 3.1. Round the quotient to the nearest hundredth.

6. Circle the number on Mr. Jefferson's bulletin board under which you would find the treasure.

**TREASURE HUNT FOR THIS WEEK**

| | | | | | | | | |
|---|---|---|---|---|---|---|---|---|
| 22.8 | | 13.75 | | 49.3 | | 182.3 | | 12.32 |
| | 7.4 | | 30 | | 2.3 | | 24 | |
| 70.28 | | 65.2 | | 3.14 | | 7.35 | | 11.8 |
| | 14.1 | | 6.28 | | 9.85 | | 6.87 | |
| 15.26 | | 31.84 | | 65.98 | | 22.25 | | 14.42 |

Answers are located on p. 105.

NAME ______________________ DATE ____________ PERIOD ______

# 5-1 Greatest Common Factor (pages 177–180)

Two or more numbers may both have the same factor, called a common factor. The greatest of the common factors of two or more numbers is called the **greatest common factor** (GCF) of the numbers. There are two methods you can use to find the GCF of two or more numbers.

| | |
|---|---|
| **Method 1: Listing Factors** | • List all of the factors of each number.<br>• Identify the common factors.<br>• The greatest of the common factors is the GCF. |
| **Method 2: Use Prime Factors** | • Write the prime factorization of each number<br>• Identify all of the common prime factors.<br>• The product of the common prime factors is the GCF. |

## EXAMPLES

**A** Find the GCF of 15 and 18.

*Make a list of the factors of each number.*

*factors of 15:* ***1***, ***3***, *5, 15*
*factors of 18:* ***1***, *2,* ***3***, *6, 9, 18*

*The common factors are 1 and 3.*
*The GCF of 15 and 18 is 3.*

**B** Find the GCF of 20 and 28.

*Write the prime factorization of each number.*

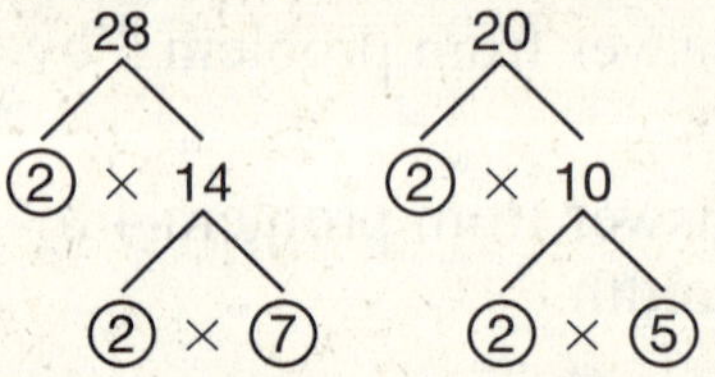


*The common prime factors are 2 and 2. The GCF of 20 and 28 is 2 × 2, or 4.*

## Try These Together

1. Find the GCF of 14 and 28.
*HINT: Make a list of factors.*

2. Find the GCF of 32 and 44.
*HINT: Use factor trees to find the common prime factors.*

## PRACTICE

***Find the GCF of each set of numbers.***

3. 7, 42
4. 10, 36
5. 44, 66
6. 30, 35
7. 4, 12, 28
8. 26, 52, 91
9. 62, 93
10. 59, 118
11. 25, 75
12. 30, 33
13. 14, 18, 22
14. 38, 57, 114

15. **Sales** Anton has made 24 gingersnaps, 60 peanut butter cookies, and 84 sugar cookies for a bake sale. What is the greatest number of boxes that he can pack them in so that the boxes contain the same number and types of cookies?

16. **Standardized Test Practice** What is the GCF of 40 and 72?

**A** 2 **B** 4 **C** 8 **D** 16

Answers: 1. 14 2. 4 3. 7 4. 2 5. 22 6. 5 7. 4 8. 13 9. 31 10. 59 11. 25 12. 3 13. 2 14. 19 15. 12 boxes 16. C

NAME ______________________ DATE ____________ PERIOD ______

# 5-2 Simplifying Fractions (pages 182–185)

You can write the fraction $\frac{2}{4}$ as $\frac{1}{2}$ and also as $\frac{4}{8}$. These fractions are **equivalent fractions,** because they name the same number. Use equivalent fractions to write fractions in **simplest form.** A fraction is in simplest form when the GCF of the numerator and denominator is 1.

| | |
|---|---|
| **Finding Equivalent Fractions** | 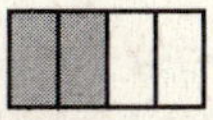 Two out of four, or $\frac{2}{4}$ of the parts of the rectangle are shaded.<br>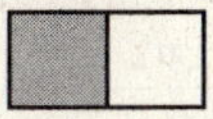 One out of two, or $\frac{1}{2}$ of the parts of the rectangle is shaded.<br>The rectangles are the same size, and the same amount of each is shaded, so the fractions are equivalent. |
| | $\frac{2}{4} \overset{\div 2}{=} \frac{1}{2}$    $\frac{1}{2} \overset{\times 2}{=} \frac{2}{4}$    Multiply or divide both the numerator and the denominator of a fraction by the same nonzero number. |

## EXAMPLES

***Replace each ■ with a number so that the fractions are equivalent.***

**A** $\frac{2}{3} = \frac{6}{■}$

*Since 2 × 3 = 6, multiply the denominator also by 3.*

$\frac{2}{3} = \frac{6}{9}$

**B** $\frac{15}{20} = \frac{■}{4}$

*Since 20 ÷ 5 = 4, divide the numerator also by 5.*

$\frac{15}{20} = \frac{3}{4}$

### Try These Together

**1.** $\frac{5}{6} = \frac{20}{■}$

*HINT: Multiply the numerator and denominator by the same number.*

**2.** Write $\frac{10}{12}$ in simplest form.

*HINT: The GCF of the numerator and denominator must be 1.*

## PRACTICE

***Replace each ■ with a number so that the fractions are equivalent.***

**3.** $\frac{2}{3} = \frac{18}{■}$    **4.** $\frac{8}{24} = \frac{■}{3}$    **5.** $\frac{5}{6} = \frac{30}{■}$

**6.** **Standardized Test Practice** What is $\frac{27}{30}$ in simplest form?

**A** $\frac{2}{3}$    **B** $\frac{9}{15}$    **C** $\frac{22}{24}$    **D** $\frac{9}{10}$

Answers: 1. 24 2. $\frac{5}{6}$ 3. 27 4. 1 5. 36 6. D

NAME ______________________ DATE ____________ PERIOD _____

# 5-3 Mixed Numbers and Improper Fractions (pages 186–189)

A **mixed number** shows the sum of a whole number and a fraction. For example, $2\frac{5}{6}$ is a mixed number that means $2 + \frac{5}{6}$. A fraction such as $\frac{8}{7}$, where the numerator is greater than or equal to the denominator, is known as an **improper fraction**. You can rewrite a mixed number as an improper fraction.

| | |
|---|---|
| **Writing Mixed Numbers as Improper Fractions** | To write a mixed number as an improper fraction, first multiply the whole number by the denominator and add the numerator. Write this sum over the denominator. $2\frac{1}{8} = \frac{(2 \times 8) + 1}{8} = \frac{17}{8}$ |
| **Writing Improper Fractions as Mixed Numbers** | Express $\frac{5}{3}$ as a mixed number. Divide the numerator by the denominator. $\begin{array}{r} 1 \\ 3\overline{)5} \\ -3 \\ \hline 2 \end{array}$ Write the remainder in the numerator of a fraction that has the divisor as the denominator. So $\frac{5}{3} = 1\frac{2}{3}$. |

## EXAMPLES

**A** Write $3\frac{2}{3}$ as an improper fraction.

$3\frac{2}{3} = \frac{(3 \times 3) + 2}{3} = \frac{11}{3}$ *Multiply 3 by 3 and add 2. Write the result over 3.*

**B** Write $\frac{8}{7}$ as a mixed number.

$8 \div 7 = 1$ R1 *Write the remainder in the numerator of a fraction that has the divisor as the denominator.*

$\frac{8}{7} = 1\frac{1}{7}$

## PRACTICE

***Write each mixed number as an improper fraction.***

**1.** $4\frac{1}{7}$ **2.** $10\frac{2}{5}$ **3.** $3\frac{1}{2}$ **4.** $5\frac{5}{9}$

***Write each improper fraction as a mixed number.***

**5.** $\frac{11}{2}$ **6.** $\frac{16}{5}$ **7.** $\frac{23}{8}$ **8.** $\frac{25}{3}$

**9. Standardized Test Practice** Write two and two-ninths as an improper fraction.

**A** $\frac{22}{9}$ **B** $\frac{20}{9}$ **C** $\frac{18}{9}$ **D** $\frac{12}{9}$

Answers: 1. $\frac{29}{7}$ 2. $\frac{52}{5}$ 3. $\frac{7}{2}$ 4. $\frac{50}{9}$ 5. $5\frac{1}{2}$ 6. $3\frac{1}{5}$ 7. $2\frac{7}{8}$ 8. $8\frac{1}{3}$ 9. B

# 5-4 Least Common Multiple (pages 194–197)

A **multiple** of a number is the product of that number and any whole number. Two different numbers can share some of the same multiples. These are called **common multiples**. The least of the common multiples of two or more numbers, other than zero, is called the **least common multiple (LCM)**. Use the following methods to find the LCM.

| | |
|---|---|
| **Method 1: Make a List** | • List the nonzero multiples of each number.<br>• Identify the LCM from the common multiples. |
| **Method 2: Use Prime Factors** | • Write the prime factorization for each number.<br>• Identify all common prime factors. Then find the product of the common prime factors using each common factor only once, and multiply by any remaining prime factors. This product is the LCM. |

## EXAMPLES

**A** Find the LCM of 4 and 6 by making a list.

*multiples of 4: 4, 8, **12**, 16, 20, **24***

*multiples of 6: 6, **12**, 18, **24**, 30*

*The LCM of 4 and 6 is 12.*

**B** Find the LCM of 10 and 12.

*Use prime factorization.*

$10 = 2 \times 5$

$12 = 2 \times 2 \times 3$

*The LCM is $2 \times 2 \times 3 \times 5$, or 60.*

## Try These Together

**1.** Find the LCM of 6 and 8.
*HINT: List the nonzero multiples of each number.*

**2.** Find the LCM of 8 and 10.
*HINT: Use prime factorization. Use common prime factors only once.*

## PRACTICE

***Find the LCM of each set of numbers.***

**3.** 2 and 7 **4.** 8 and 12 **5.** 25 and 30 **6.** 6 and 21
**7.** 3 and 8 **8.** 8 and 18 **9.** 4 and 10 **10.** 15 and 35
**11.** 7 and 14 **12.** 3 and 5 **13.** 4 and 9 **14.** 4 and 22
**15.** 20 and 45 **16.** 2, 9, and 15 **17.** 3, 15, and 45 **18.** 10, 30, and 65

**19. Design** Ingrid is stringing 3 bracelets, one with 4 mm beads, one with 5 mm beads, and one with 6 mm beads. What is the shortest length where all the bracelets are equal?

**20. Standardized Test Practice** Find the LCM of 5, 6, and 45.

**A** 45 **B** 60 **C** 90 **D** 135

Answers: **1.** 24 **2.** 40 **3.** 14 **4.** 24 **5.** 150 **6.** 42 **7.** 24 **8.** 72 **9.** 20 **10.** 105 **11.** 14 **12.** 15 **13.** 36 **14.** 44 **15.** 180 **16.** 90 **17.** 45 **18.** 390 **19.** 60 mm **20.** C

NAME ______________________ DATE ____________ PERIOD _____

# 5-5 Comparing and Ordering Fractions

(pages 198–201)

To compare fractions with different denominators, find the **least common denominator (LCD)**, or the LCM of the denominators.

## EXAMPLES

**A** Find the LCD for $\frac{1}{2}$ and $\frac{1}{3}$.

*The LCD of $\frac{1}{2}$ and $\frac{1}{3}$ is the LCM of 2 and 3.*

*Multiples of 2: 0, 2, 4, **6**, 8*
*Multiples of 3: 0, 3, **6**, 9*

*The LCM of 2 and 3 is 6, so the LCD for $\frac{1}{2}$ and $\frac{1}{3}$ is also 6.*

**B** Which fraction is greater, $\frac{2}{3}$ or $\frac{3}{4}$?

*Find the LCD of $\frac{2}{3}$ and $\frac{3}{4}$. The LCM of 3 and 4 is 12, so the LCD is also 12.*

*$\frac{2}{3} = \frac{8}{12}$ and $\frac{3}{4} = \frac{9}{12}$. Multiply the numerator and denominator of $\frac{2}{3}$ by 4 and multiply the numerator and denominator of $\frac{3}{4}$ by 3 in order to rewrite $\frac{2}{3}$ and $\frac{3}{4}$ as equivalent fractions with 12 as the denominator. Since $\frac{8}{12} < \frac{9}{12}$, it is true that $\frac{2}{3} < \frac{3}{4}$, so $\frac{3}{4}$ is the greater fraction.*

## Try These Together

**1.** Find the LCD for $\frac{2}{5}$ and $\frac{1}{6}$.

*HINT: Find the LCM of the denominators.*

**2.** Which fraction is greater, $\frac{1}{4}$ or $\frac{2}{5}$?

*HINT: Find the LCD and then multiply both numerator and denominator to rewrite the fractions with the same denominator.*

## PRACTICE

### *Find the LCD for each pair of fractions.*

**3.** $\frac{2}{5}, \frac{1}{3}$ **4.** $\frac{4}{7}, \frac{9}{14}$ **5.** $\frac{3}{10}, \frac{7}{8}$ **6.** $\frac{1}{4}, \frac{3}{8}$

### *Replace each ● with <, >, or = to make a true statement.*

**7.** $\frac{4}{7}$ ● $\frac{8}{14}$ **8.** $\frac{2}{7}$ ● $\frac{1}{9}$ **9.** $\frac{1}{6}$ ● $\frac{3}{18}$ **10.** $\frac{2}{5}$ ● $\frac{1}{3}$

**11.** $\frac{1}{5}$ ● $\frac{2}{10}$ **12.** $\frac{4}{34}$ ● $\frac{3}{17}$ **13.** $\frac{11}{12}$ ● $\frac{13}{16}$ **14.** $\frac{13}{22}$ ● $\frac{7}{11}$

**15. Population** The U.S. Census Bureau estimates that 10- to 19-year-olds are about $\frac{3}{20}$ of the population, and 35- to 44-year-olds are about $\frac{4}{25}$. Which age group represents more of the population?

**16. Standardized Test Practice** Order the fractions $\frac{1}{7}$, $\frac{2}{6}$, and $\frac{3}{8}$ from least to greatest.

**A** $\frac{3}{8}, \frac{2}{6}, \frac{1}{7}$ **B** $\frac{1}{7}, \frac{3}{8}, \frac{2}{6}$ **C** $\frac{2}{6}, \frac{1}{7}, \frac{3}{8}$ **D** $\frac{1}{7}, \frac{2}{6}, \frac{3}{8}$

Answers: 1. 30 2. $\frac{2}{5}$ 3. 15 4. 14 5. 40 6. 8 7. = 8. > 9. = 10. > 11. = 12. < 13. > 14. < 15. 35–44 16. D

NAME ______________________ DATE ____________ PERIOD _____

# 5-6 Writing Decimals as Fractions

(pages 202–205)

Decimals like 0.58, 0.32, 0.16, and 0.08 can be written as fractions with denominators of 10, 100, 1,000, and so on.

## EXAMPLES

**A** Write 0.5 as a fraction in simplest form.

| | |
|---|---|
| 0.5 | *The decimal 0.5 is read as "five tenths."* |
| $0.5 = \frac{5}{10}$ | *Write the decimal as the fraction "five tenths."* |
| $= \frac{1}{2}$ | *Simplify. Divide the numerator and the denominator each by the GCF, 5.* |

**B** Write 2.25 as a mixed number in simplest form.

| | |
|---|---|
| 2.25 | *The decimal is read as "two and twenty-five hundredths."* |
| $2.25 = 2\frac{25}{100}$ | *Write the decimal as the mixed number "two and twenty-five hundredths."* |
| $= 2\frac{1}{4}$ | *Simplify. Divide the numerator and the denominator each by the GCF, 25.* |

### Try These Together

***Write each decimal as a fraction or mixed number in simplest form.***

**1.** 0.62

*HINT: Say the decimal aloud, and then write it as a fraction. Simplify the fraction.*

**2.** 12.84

*HINT: Say the decimal aloud and then write it as a mixed number. Simplify the mixed number.*

## PRACTICE

***Write each decimal as a fraction or mixed number in simplest form.***

**3.** 3.3  **4.** 2.15  **5.** 4.007  **6.** 1.78

**7.** 7.66  **8.** 4.1  **9.** 7.91  **10.** 8.02

**11.** 3.8  **12.** 0.08  **13.** 9.76  **14.** 4.03

**15.** 5.25  **16.** 0.034  **17.** 9.28  **18.** 3.48

**19. Fashion** A bottle of hairspray holds 8.45 fluid ounces. Express this as a mixed number in simplest form.

**20. Standardized Test Practice** Write two and forty-four hundredths as a mixed number in simplest form.

**A** $2\frac{11}{25}$  **B** $2\frac{44}{100}$  **C** $2\frac{11}{250}$  **D** $2\frac{22}{50}$

Answers: **1.** $\frac{31}{50}$ **2.** $12\frac{21}{25}$ **3.** $3\frac{3}{10}$ **4.** $2\frac{3}{20}$ **5.** $4\frac{7}{1,000}$ **6.** $1\frac{39}{50}$ **7.** $7\frac{33}{50}$ **8.** $4\frac{1}{10}$ **9.** $7\frac{91}{100}$ **10.** $8\frac{1}{50}$ **11.** $3\frac{4}{5}$ **12.** $\frac{2}{25}$ **13.** $9\frac{19}{25}$ **14.** $4\frac{3}{100}$ **15.** $5\frac{1}{4}$ **16.** $\frac{17}{500}$ **17.** $9\frac{7}{25}$ **18.** $3\frac{12}{25}$ **19.** $8\frac{9}{20}$ **20.** A

NAME ____________________ DATE ____________ PERIOD ______

# 5-7 Writing Fractions as Decimals

(pages 206–209)

Any fraction can be written as a decimal by using division.

| | |
|---|---|
| **Terminating Decimals** | Decimals like 0.45 and 0.85 are **terminating decimals** because the division ends, or terminates, when the remainder is zero. $\frac{4}{5}$ means $4 \div 5$. Divide 4 by 5, and the quotient 0.8 is the decimal you want to find. |
| **Repeating Decimals** | Decimals like 0.333333 . . . are called **repeating decimals** because the digits repeat. **Bar notation** can be used to indicate that decimals repeat. $0.6666666\ldots = 0.\overline{6}$, $0.277777\ldots = 0.2\overline{7}$, $0.737373\ldots = 0.\overline{73}$ Bar notation is useful because some fractions, when written as decimals, are repeating decimals. For example, $\frac{2}{3} = 0.\overline{6}$. |

## EXAMPLES

***Write each fraction as a decimal.***

**A** $\frac{1}{5}$

*$\frac{1}{5} = 1 \div 5$*

```
  0.2
5)1.0      Divide 1 by 5.
 -1 0
    0      Therefore, 1/5 = 0.2.
```

*Therefore, $\frac{1}{5} = 0.2$.*

**B** $\frac{1}{3}$

*$\frac{1}{3} = 1 \div 3$*

```
  0.33
3)1.00     Divide 1 by 3.
  -9
   10      This pattern will continue forever.
   -9
   10      1/3 is a repeating decimal, 0.3̄.
```

*This pattern will continue forever.*

*$\frac{1}{3}$ is a repeating decimal, $0.\overline{3}$.*

### Try These Together

***Write each fraction or mixed number as a decimal.***

**1.** $\frac{3}{4}$ *HINT: Divide 3 by 4.*

**2.** $2\frac{1}{2}$ *HINT: The whole number is written to the left of the decimal point.*

## PRACTICE

***Write each fraction or mixed number as a decimal.***

**3.** $4\frac{1}{8}$ **4.** $\frac{1}{6}$ **5.** $\frac{5}{9}$ **6.** $\frac{2}{5}$

**7.** $5\frac{11}{12}$ **8.** $\frac{8}{11}$ **9.** $\frac{8}{9}$ **10.** $6\frac{3}{10}$

**11.** **Standardized Test Practice** Write $2\frac{5}{12}$ as a decimal.

**A** 2.4166 **B** $2.41\overline{6}$ **C** $2.\overline{146}$ **D** 2.41666

Answers: **1.** 0.75 **2.** 2.5 **3.** 4.125 **4.** $0.1\overline{6}$ **5.** $0.\overline{5}$ **6.** 0.4 **7.** $5.91\overline{6}$ **8.** $0.\overline{72}$ **9.** $0.\overline{8}$ **10.** 6.3 **11.** B

NAME ________________________ DATE ____________ PERIOD ______

# 5 Chapter 5 Review

## Funny Money

Until recently, the prices of stocks sold on the New York Stock Exchange were listed as mixed numbers. For example, the price of a stock would be $\$58\frac{1}{4}$ instead of \$58.25.

When you go to the corner store, you see prices displayed in dollars and cents, or in decimal form. Suppose you go to the corner store one day, and you see all of the prices displayed as fractions and mixed numbers. Will you know how much to pay?

**1.** You go to the cooler for a soda. The price of the bottle is listed as $\frac{4}{5}$ of a dollar. What is this price in dollars and cents?

0.80

**2.** You see a sign saying granola bars are on sale. The price is $\$1\frac{2}{8}$. If a candy bar costs $\$1\frac{1}{5}$, which bar is less expensive? How much is each bar in dollars and cents?

Less expensive - candy bar
Candy bar = 1.20 ¢
Granola bar = 1.25

**3.** Draw lines to match the prices of the items in the left column with the prices in the right column. All prices have been rounded to the nearest cent.

| | | |
|---|---|---|
| banana (1) | $\$\frac{1}{8}$ | \$1.40 |
| paper towel (roll) | $\$1\frac{2}{5}$ | \$0.30 |
| one dozen eggs | $\$\frac{19}{20}$ | \$0.13 |
| hard candies (each) | $\$\frac{3}{10}$ | \$0.95 |

**4.** One of your favorite snacks, bagels, used to sell for \$1.33 each. What would they sell for now that the store uses fractional prices?

Answers are located on p. 105.

NAME ______________________ DATE ____________ PERIOD _____

# 6-1 Rounding Fractions and Mixed Numbers (pages 219–222)

The following guidelines can help you round fractions and mixed numbers to the nearest unit.

| | |
|---|---|
| **Rounding Fractions and Mixed Numbers** | • If the numerator is almost as large as the denominator, round the number up to the next whole number.<br>• If the numerator is about half of the denominator, round the fraction to $\frac{1}{2}$.<br>• If the numerator is much smaller than the denominator, round the number down to the next whole number.<br>• When measuring actual quantities, you may have to round up or down, despite what the rule says, to get useful numbers. |

## EXAMPLES

**A** Round $\frac{7}{8}$ to the nearest half.

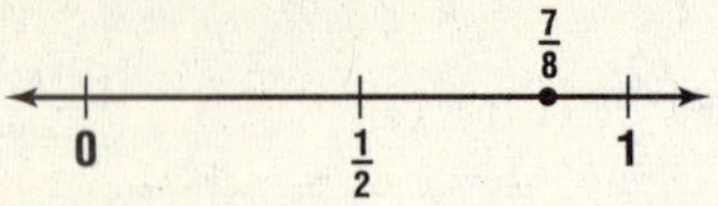


*The numerator is almost as large as the denominator, so round up. Since $\frac{7}{8}$ is closer to 1 than $\frac{1}{2}$, round up to 1.*

**B** Round $3\frac{3}{5}$ to the nearest half.

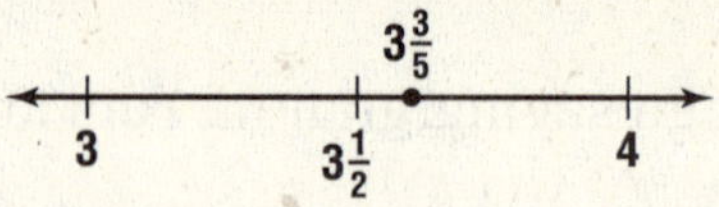


*The numerator is about half of the denominator. Round the fraction to $\frac{1}{2}$. So, $3\frac{3}{5}$ rounds to $3\frac{1}{2}$.*

### Try These Together

***Round each number to the nearest half.***

1. $\frac{2}{5}$ *HINT: The numerator is about half of the denominator.*

2. $5\frac{1}{8}$ *HINT: The numerator is much smaller than the denominator.*

## PRACTICE

***Round each number to the nearest half.***

3. $1\frac{5}{8}$
4. $\frac{7}{12}$
5. $2\frac{3}{8}$
6. $8\frac{9}{10}$
7. $\frac{4}{6}$
8. $1\frac{2}{9}$

9. **Standardized Test Practice** A hot air balloon can carry 400 pounds of cargo and people. There are four men who want to ride in the balloon. The average weight of the men is 180 pounds. Estimate how many men can ride in the balloon.

**A** 4 **B** 2 **C** 3 **D** 1

Answers: 1. $\frac{1}{2}$ 2. 5 3. $1\frac{1}{2}$ 4. $\frac{1}{2}$ 5. $2\frac{1}{2}$ 6. 9 7. $\frac{1}{2}$ 8. 1 9. B

NAME ______________________ DATE ____________ PERIOD _____

# 6-2 Estimating Sums and Differences

(pages 223–225)

When you add or subtract fractions or mixed numbers, round to estimate the sum or difference.

| | |
|---|---|
| **Estimate the Sum or Difference of Fractions** | • Round each fraction to the nearest half, and then add or subtract. |
| **Estimate the Sum or Difference of Mixed Numbers** | • Round each mixed number to the nearest whole number, and then add or subtract. |

## EXAMPLES

**A** Estimate $\frac{13}{15} + \frac{9}{16}$.

*$\frac{13}{15}$ rounds to 1 and $\frac{9}{16}$ rounds to $\frac{1}{2}$.*

*Add $1 + \frac{1}{2} = 1\frac{1}{2}$.*

*$\frac{13}{15} + \frac{9}{16}$ is about $1\frac{1}{2}$.*

**B** Estimate $5\frac{7}{8} - 2\frac{2}{5}$.

*$5\frac{7}{8}$ rounds to 6 and $2\frac{2}{5}$ rounds to 2.*

*Subtract $6 - 2 = 4$.*

*$5\frac{7}{8} - 2\frac{2}{5}$ is about 4.*

## Try These Together

***Estimate.***

**1.** $\frac{7}{12} - \frac{1}{7}$

*HINT: Round to the nearest half.*

**2.** $5\frac{1}{8} + 9\frac{3}{5}$

*HINT: Round to the nearest whole number.*

## PRACTICE

***Estimate.***

**3.** $\frac{2}{3} + \frac{4}{5}$

**4.** $3\frac{1}{4} - 2\frac{1}{8}$

**5.** $\frac{5}{8} + \frac{5}{6}$

**6.** $8\frac{3}{4} - 1\frac{3}{16}$

**7.** $\frac{1}{10} + \frac{4}{9}$

**8.** $\frac{2}{5} + \frac{3}{8}$

**9.** $1\frac{1}{7} - \frac{9}{10}$

**10.** $1\frac{3}{5} - \frac{1}{5}$

**11.** Estimate the sum $2\frac{1}{3} + \frac{5}{6} + 3\frac{3}{4} + 6\frac{1}{9}$.

**12.** Estimate the difference between $4\frac{1}{5}$ and $3\frac{2}{3}$.

**13.** **Standardized Test Practice** Estimate the following total.

$$\left(\frac{3}{4} + 2\frac{1}{8}\right) - \left(1\frac{1}{5} + 1\frac{3}{5}\right)$$

**A** 0 **B** $\frac{1}{2}$ **C** 1 **D** $1\frac{1}{2}$

**Answers:** Sample answers are given. **1.** $\frac{1}{2}$ **2.** 15 **3.** $1\frac{1}{2}$ **4.** 1 **5.** $1\frac{1}{2}$ **6.** 8 **7.** $\frac{1}{2}$ **8.** 1 **9.** 0 **10.** 2 **11.** 13 **12.** 0 **13.** A

NAME ______________________ DATE ____________ PERIOD ______

# 6-3 Adding and Subtracting Fractions with Like Denominators (pages 228–231)

Fractions with the same denominator are **like fractions**. You add and subtract the numerators of like fractions the same way you add and subtract whole numbers.

| | |
|---|---|
| **Adding Like Fractions** | • To add fractions with like denominators, add the numerators. Use the same denominator in the sum. |
| **Subtracting Like Fractions** | • To subtract fractions with like denominators, subtract the numerators. Use the same denominator in the difference. |

## EXAMPLES

**A** Find the sum of $\frac{1}{7}$ and $\frac{3}{7}$.

*Estimate.* $0 + \frac{1}{2} = \frac{1}{2}$

$\frac{1}{7} + \frac{3}{7} = \frac{1+3}{7}$

$= \frac{4}{7}$ *Compared to the estimate, the answer is reasonable.*

**B** Find the difference $\frac{3}{4} - \frac{1}{4}$.

*Estimate.* $1 - \frac{1}{2} = \frac{1}{2}$

$\frac{3}{4} - \frac{1}{4} = \frac{3-1}{4}$

$= \frac{2}{4}$ or $\frac{1}{2}$ *Compared to the estimate, the answer is reasonable.*

### Try These Together

***Add or subtract. Write in simplest form.***

**1.** $\frac{2}{3} + \frac{2}{3}$

*HINT: Add the numerators. Write the sum as a mixed number.*

**2.** $\frac{5}{8} - \frac{3}{8}$

*HINT: Subtract the numerators. Write the answer in simplest form.*

## PRACTICE

***Add or subtract. Write in simplest form.***

**3.** $\frac{1}{3} + \frac{2}{3}$

**4.** $\frac{4}{5} + \frac{2}{5}$

**5.** $\frac{7}{16} - \frac{3}{16}$

**6.** $\frac{9}{10} - \frac{3}{10}$

**7.** $\frac{2}{7} + \frac{3}{7}$

**8.** $\frac{9}{15} - \frac{6}{15}$

**9.** How much larger is $\frac{7}{8}$ than $\frac{3}{8}$?

**10.** Find the sum of $\frac{1}{8}$, $\frac{3}{8}$, and $\frac{5}{8}$.

**11. Standardized Test Practice** Find the following total. $\left(\frac{11}{16} + \frac{5}{16}\right) - \left(\frac{3}{16} + \frac{8}{16}\right)$

**A** $\frac{7}{16}$ **B** $\frac{1}{2}$ **C** $\frac{5}{16}$ **D** $1\frac{3}{16}$

Answers: **1.** $1\frac{1}{3}$ **2.** $\frac{1}{4}$ **3.** 1 **4.** $1\frac{1}{5}$ **5.** $\frac{1}{4}$ **6.** $\frac{3}{5}$ **7.** $\frac{5}{7}$ **8.** $\frac{1}{5}$ **9.** $\frac{1}{2}$ **10.** $1\frac{1}{8}$ **11.** C

NAME ______________________ DATE ____________ PERIOD _____

# 6-4 Adding and Subtracting Fractions with Unlike Denominators (pages 235–238)

When you add or subtract fractions, the fractions must have the same denominators. To add or subtract fractions with unlike denominators, rename the fractions using the least common denominator (LCD). Then add or subtract and simplify.

**EXAMPLES**

**A** Add $\frac{1}{2}$ and $\frac{2}{3}$.

*The LCD of $\frac{1}{2}$ and $\frac{2}{3}$ is 6.*

$\frac{1}{2} = \frac{3}{6}$ (×3), and $\frac{2}{3} = \frac{4}{6}$ (×2) — *Rename the fractions.*

$\frac{3}{6} + \frac{4}{6} = \frac{7}{6}$, or $1\frac{1}{6}$ — *Add, then simplify.*

**B** Find $\frac{3}{5} - \frac{1}{4}$.

*The LCD of $\frac{3}{5}$ and $\frac{1}{4}$ is 20.*

$\frac{3}{5} = \frac{12}{20}$ (×4), and $\frac{1}{4} = \frac{5}{20}$ (×5) — *Rename the fractions.*

$\frac{12}{20} - \frac{5}{20} = \frac{7}{20}$ — *Subtract.*

## Try These Together

***Add or subtract. Write in simplest form.***

**1.** $\frac{3}{4} - \frac{1}{6}$

*HINT: Find the LCD, then rename the fractions.*

**2.** $\frac{3}{8} + \frac{5}{12}$

*HINT: Find the LCD, then rename the fractions.*

**PRACTICE**

***Add or subtract. Write in simplest form.***

**3.** $\frac{3}{8} + \frac{1}{4}$

**4.** $\frac{2}{3} + \frac{1}{6}$

**5.** $\frac{7}{8} + \frac{1}{2}$

**6.** $\frac{2}{5} + \frac{1}{3}$

**7.** $\frac{11}{12} + \frac{5}{6}$

**8.** $\frac{1}{6} + \frac{3}{4}$

**9.** $\frac{3}{7} + \frac{1}{2}$

**10.** $\frac{8}{11} - \frac{2}{3}$

**11.** $\frac{4}{9} - \frac{1}{6}$

**12.** What is the sum of $\frac{5}{8}$ and $\frac{9}{16}$?

**13.** How much is $\frac{9}{10} - \frac{2}{5}$?

**14.** How much more is $\frac{11}{16}$ than $\frac{1}{4}$?

**15. Carpentry** You are building a bookcase. The board that makes up the side of the bookcase is $\frac{1}{2}$ inch thick. If you use $\frac{7}{8}$-inch screws to attach the shelves of the bookcase, how far into the shelves do the screws extend?

**16. Standardized Test Practice** What is the sum of $\frac{1}{6}$, $\frac{3}{4}$, and $\frac{9}{12}$?

**A** $\frac{7}{12}$ **B** $\frac{11}{12}$ **C** $1\frac{5}{12}$ **D** $1\frac{2}{3}$

Answers: 1. $\frac{7}{12}$ 2. $\frac{19}{24}$ 3. $\frac{5}{8}$ 4. $\frac{5}{6}$ 5. $1\frac{3}{8}$ 6. $\frac{11}{15}$ 7. $1\frac{3}{4}$ 8. $\frac{11}{12}$ 9. $\frac{13}{14}$ 10. $\frac{2}{33}$ 11. $\frac{5}{18}$ 12. $1\frac{3}{16}$ 13. $\frac{1}{2}$ 14. $\frac{7}{16}$ 15. $\frac{3}{8}$ inch 16. D

NAME ______________________ DATE __________ PERIOD _____

# 6-5 Adding and Subtracting Mixed Numbers (pages 240–243)

Use the following rules to add and subtract mixed numbers.

| | |
|---|---|
| **Adding and Subtracting Mixed Numbers** | • Add or subtract the fractions.<br>• Then add or subtract the whole numbers.<br>• Rename and simplify if necessary. |

## EXAMPLES

**A** Find $5\frac{5}{8} + 1\frac{1}{8}$.

*Add the fractions.* *Add the whole numbers.*

$$\begin{array}{r} 5\frac{5}{8} \\ +1\frac{1}{8} \\ \hline \frac{6}{8} \end{array} \rightarrow \begin{array}{r} 5\frac{5}{8} \\ +1\frac{1}{8} \\ \hline 6\frac{6}{8} \end{array} \text{ or } 6\frac{3}{4} \text{ Simplify.}$$

**B** Find $3\frac{5}{6} - 2\frac{1}{2}$.

*Subtract the fractions.* *Subtract the whole numbers.*

$$\begin{array}{r} 3\frac{5}{6} \\ -2\frac{1}{2} \\ \hline \end{array} \rightarrow \begin{array}{r} 3\frac{5}{6} \\ -2\frac{3}{6} \\ \hline \frac{2}{6} \end{array} \rightarrow \begin{array}{r} 3\frac{5}{6} \\ -2\frac{3}{6} \\ \hline 1\frac{2}{6} \end{array} \text{ or } 1\frac{1}{3} \text{ Simplify.}$$

### Try These Together

***Add or subtract. Write in simplest form.***

**1.** $7\frac{1}{4} + 10\frac{1}{2}$

*HINT: Rename the fractions. Add the fractions. Then add the whole numbers.*

**2.** $9\frac{11}{12} - 4\frac{3}{8}$

*HINT: Rename the fractions. Subtract the fractions. Then subtract the whole numbers.*

## PRACTICE

***Add or subtract. Write in simplest form.***

**3.** $2\frac{1}{3} + 5\frac{3}{8}$ **4.** $9\frac{3}{5} - 2\frac{3}{15}$ **5.** $5\frac{2}{3} + 3\frac{1}{2}$

**6.** $8\frac{1}{3} - 6\frac{1}{4}$ **7.** $15\frac{7}{8} - 12\frac{3}{4}$ **8.** $8\frac{5}{12} - 2\frac{1}{8}$

**9.** $1\frac{7}{10} + 4\frac{1}{3}$ **10.** $9\frac{1}{3} + 5\frac{5}{6}$ **11.** $4\frac{3}{4} - 2\frac{2}{3}$

**12. Standardized Test Practice** A bag of potatoes weighs $5\frac{3}{4}$ pounds. At the first meal, $1\frac{1}{3}$ pounds of potatoes are eaten. At a later meal, $2\frac{1}{4}$ pounds of potatoes are eaten. How many pounds of potatoes remain in the bag?

**A** $2\frac{1}{6}$ **B** $2\frac{1}{3}$ **C** $1\frac{5}{6}$ **D** $2\frac{2}{3}$

Answer: **1.** $17\frac{3}{4}$ **2.** $5\frac{13}{24}$ **3.** $7\frac{17}{24}$ **4.** $7\frac{2}{5}$ **5.** $9\frac{1}{6}$ **6.** $2\frac{1}{12}$ **7.** $3\frac{1}{8}$ **8.** $6\frac{7}{24}$ **9.** $6\frac{1}{30}$ **10.** $15\frac{1}{6}$ **11.** $2\frac{1}{12}$ **12.** A

NAME ______________________________ DATE ____________ PERIOD _____

# 6-6 Subtracting Mixed Numbers with Renaming (pages 244–247)

When you subtract mixed numbers, sometimes the fraction in the number you are subtracting is greater than the fraction in the number you are subtracting from. When this happens, you must rename the first fraction as an improper fraction in order to subtract.

## EXAMPLE

Find $12\frac{1}{3} - 8\frac{3}{5}$.

*The LCM of 3 and 5 is 15.*

$$\begin{array}{r} 12\frac{1}{3} \\ -8\frac{3}{5} \\ \hline \end{array} \rightarrow \begin{array}{r} 12\frac{5}{15} \\ -8\frac{9}{15} \\ \hline \end{array}$$

*Since $\frac{9}{15}$ is greater than $\frac{5}{15}$, rename $12\frac{5}{15}$ as $11\frac{20}{15}$, and then subtract.*

$$\begin{array}{r} 11\frac{20}{15} \\ -8\frac{9}{15} \\ \hline \end{array} \rightarrow \begin{array}{r} 11\frac{20}{15} \\ -8\frac{9}{15} \\ \hline \frac{11}{15} \end{array} \rightarrow \begin{array}{r} 11\frac{20}{15} \\ -8\frac{9}{15} \\ \hline 3\frac{11}{15} \end{array}$$

### Try These Together

***Subtract. Write in simplest form.***

**1.** $4\frac{3}{10} - 1\frac{7}{10}$

*HINT: Rename the fraction in the first mixed number.*

**2.** $8\frac{3}{14} - 5\frac{5}{7}$

*HINT: First find the LCD. Then rename using the LCD. Then, rename the first fraction as an improper fraction.*

## PRACTICE

***Subtract. Write in simplest form.***

**3.** $9\frac{5}{12} - 4\frac{11}{12}$ **4.** $4\frac{3}{10} - 2\frac{9}{10}$ **5.** $7\frac{2}{5} - 6\frac{3}{10}$

**6.** $18\frac{3}{4} - 7\frac{7}{8}$ **7.** $5\frac{3}{8} - 1\frac{5}{8}$ **8.** $9\frac{1}{6} - 7\frac{2}{5}$

**9.** $4\frac{3}{4} - 3\frac{1}{8}$ **10.** $3\frac{1}{2} - 1\frac{5}{8}$ **11.** $18 - \frac{5}{6}$

**12. Algebra** Solve the equation $m = 9\frac{5}{8} - 6\frac{4}{5}$. Write the solution in simplest form.

**13. Standardized Test Practice** Sam swam $2\frac{3}{8}$ hours on Saturday and $3\frac{1}{3}$ hours on Sunday. How many more hours did he swim on Sunday than on Saturday?

**A** $1\frac{1}{24}$ **B** $\frac{23}{24}$ **C** $1\frac{3}{8}$ **D** $1\frac{1}{4}$

Answers: **1.** $2\frac{3}{5}$ **2.** $2\frac{1}{2}$ **3.** $4\frac{1}{2}$ **4.** $1\frac{2}{5}$ **5.** $1\frac{1}{10}$ **6.** $10\frac{7}{8}$ **7.** $3\frac{3}{4}$ **8.** $1\frac{23}{30}$ **9.** $1\frac{5}{8}$ **10.** $1\frac{7}{8}$ **11.** $17\frac{1}{6}$ **12.** $2\frac{33}{40}$ **13.** B

# 6 Chapter 6 Review

## Pie-Eating Contest

You've just entered a pie-eating contest, but this contest is a little different from most. You must eat the correct amount of pie in each round of the contest to win. The instructions for each round tell you how much pie to eat. Shade the blank pies below to show how much pie you would eat in each round.

**Round 1:** Eat $\frac{2}{5} + \frac{1}{5}$ of the pie.

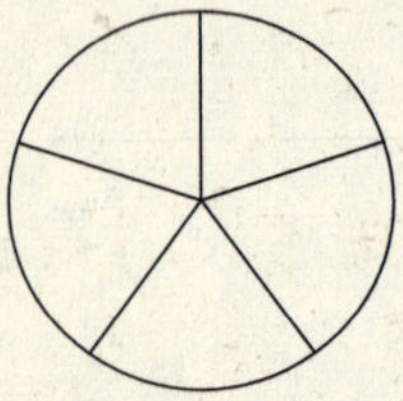

**Round 2:** Eat $\frac{2}{3} - \frac{1}{4}$ of the pie.

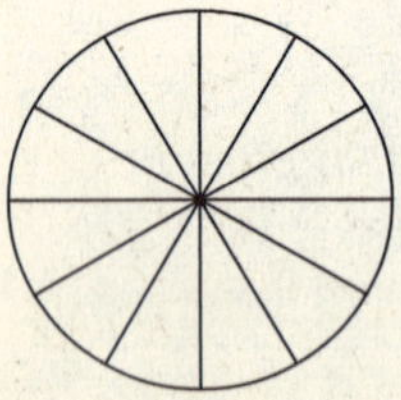

**Round 3:** Eat $1\frac{1}{2} + 1\frac{3}{8}$ of the pies.

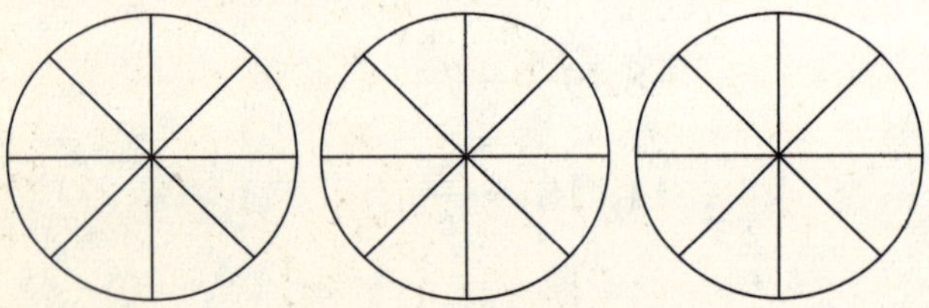

**Round 4:** Eat $2\frac{1}{4} - 1\frac{7}{8}$ of the pies.

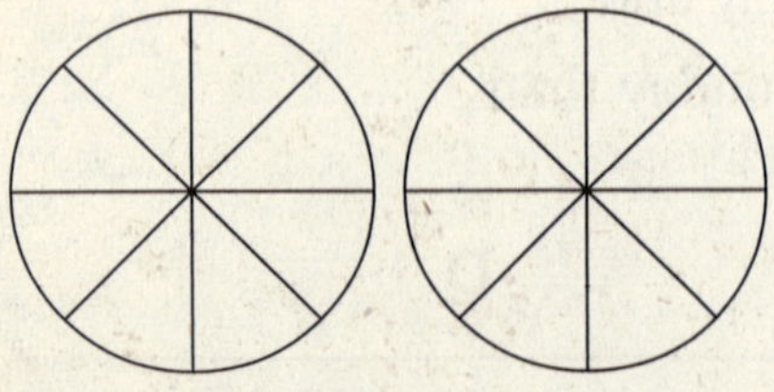

Answers are located on p. 106.

NAME ____________________ DATE ____________ PERIOD _____

# 7-1 Estimating Products (pages 256–258)

You can use **compatible numbers** to estimate products when multiplying fractions. Compatible numbers are easy to divide mentally.

**EXAMPLES**

**A** Estimate $\frac{1}{4} \times 13$.

*$\frac{1}{4} \times 13$ means $\frac{1}{4}$ of 13.*

*$\frac{1}{4} \times 12 = \underline{\ ?\ }$ For 13, the nearest multiple of 4 is 12. 4 and 12 are compatible numbers because $12 \div 4 = 3$.*

*$\frac{1}{4} \times 12 = 3$, so the product of $\frac{1}{4}$ and 13 is about 3.*

**B** Estimate $\frac{2}{3} \times 17$.

*$\frac{1}{3} \times 18 = 6$ For 17, the nearest multiple of 3 is 18. $\frac{1}{3}$ of 18 is 6.*

*$\frac{2}{3} \times 18 = 12$ Since $\frac{1}{3}$ of 18 is 6, it follows that $\frac{2}{3}$ of 18 is $2 \times 6$ or 12.*

*So, $\frac{2}{3} \times 17$ is about 12.*

You can also estimate products by rounding fractions to 0, $\frac{1}{2}$, or 1, and by rounding mixed numbers to the nearest whole numbers.

## Try These Together

***Estimate each product.***

1. Estimate $\frac{1}{5} \times 9$.
   *HINT: For 9, what is the nearest multiple of 5?*
2. Estimate $\frac{5}{6} \times 22$.
   *HINT: For 22, what is the nearest multiple of 6?*

**PRACTICE**

***Estimate each product.***

3. $\frac{1}{5} \times 24$
4. $\frac{1}{6} \times 5$
5. $\frac{5}{8} \times 42$
6. $2\frac{1}{4} \times 3\frac{1}{3}$
7. $\frac{1}{10} \times \frac{5}{8}$
8. $6\frac{2}{3} \times 1\frac{4}{5}$
9. $\frac{4}{9} \times 14$
10. $3\frac{4}{5} \times 7\frac{1}{8}$
11. $4\frac{7}{9} \times 2\frac{1}{6}$

12. **Standardized Test Practice** Ann receives an allowance of $10 a week. She spends about $\frac{2}{3}$ of her allowance on school lunches and about $\frac{1}{6}$ on entertainment. About how much does she have left?

**A** $2 **B** $0 **C** $8 **D** $1

**Answers:** Sample answers are given. **1.** 2 **2.** 20 **3.** 5 **4.** 1 **5.** 25 **6.** 6 **7.** 0 **8.** 14 **9.** 7 **10.** 28 **11.** 10 **12.** A

NAME ______________________ DATE ____________ PERIOD ______

# 7-2 Multiplying Fractions (pages 261–264)

Use the following rules to multiply fractions.

| | |
|---|---|
| **Multiplying Fractions** | To multiply fractions, multiply the numerators and multiply the denominators. Simplify if necessary. |
| **Simplify Before You Multiply** | You can simplify before you multiply fractions if the numerator of one fraction and the denominator of another fraction have a common factor. |

## EXAMPLES

***Multiply.***

**A** $\frac{1}{3} \times \frac{2}{5}$

$\frac{1}{3} \times \frac{2}{5} = \frac{1 \times 2}{3 \times 5}$ *To multiply fractions, multiply the numerators and the denominators*

$= \frac{2}{15}$ *You cannot simplify $\frac{2}{15}$.*

**B** $\frac{4}{7} \times \frac{3}{8}$

$\frac{4}{7} \times \frac{3}{8}$ *Estimate: $\frac{1}{2} \times \frac{1}{2} = \frac{1}{4}$*

$= \frac{\cancel{4}^{1} \times 3}{7 \times \cancel{8}_{2}}$ *The GCF of 4 and 8 is 4. Divide both the numerator and denominator by 4 and then multiply.*

$= \frac{3}{14}$

## Try These Together

***Multiply.***

**1.** $\frac{1}{2} \times \frac{3}{8}$

*HINT: Multiply the numerators and the denominators.*

**2.** $\frac{5}{6} \times \frac{3}{25}$

*HINT: Simplify before you multiply.*

## PRACTICE

***Multiply. Write in simplest form.***

**3.** $\frac{1}{2} \times \frac{3}{4}$ **4.** $\frac{5}{8} \times \frac{2}{3}$ **5.** $\frac{2}{3} \times \frac{6}{8}$

**6.** $\frac{2}{3} \times \frac{1}{9}$ **7.** $\frac{3}{5} \times \frac{5}{12}$ **8.** $\frac{1}{3} \times \frac{9}{10}$

**9.** $\frac{1}{12} \times \frac{4}{5}$ **10.** $\frac{3}{7} \times \frac{4}{9}$ **11.** $\frac{3}{5} \times \frac{3}{4}$

**12. Standardized Test Practice** There are a dozen eggs in a carton. You use $\frac{1}{6}$ for an omelet. Your sister uses $\frac{1}{5}$ of the leftover eggs for a cake. How many eggs are left?

**A** 10 **B** 2 **C** 8 **D** 6

Answers: 1. $\frac{3}{16}$ 2. $\frac{1}{10}$ 3. $\frac{3}{8}$ 4. $\frac{5}{12}$ 5. $\frac{1}{2}$ 6. $\frac{2}{27}$ 7. $\frac{1}{4}$ 8. $\frac{3}{10}$ 9. $\frac{1}{15}$ 10. $\frac{4}{21}$ 11. $\frac{9}{20}$ 12. C

NAME ______________________ DATE ____________ PERIOD _____

# 7-3 Multiplying Mixed Numbers

(pages 265–267)

Use the following rules to multiply mixed numbers.

| | |
|---|---|
| **Multiplying Mixed Numbers** | • Express mixed numbers as improper fractions.<br>• Multiply the numerators and multiply the denominators. |
| **Simplify Before You Multiply** | After you express mixed numbers as improper fractions, check to see if the numerator of one fraction and the denominator of another fraction have a common factor. If they do, simplify before you multiply. |

## EXAMPLES

***Multiply.***

**A** $1\frac{1}{4} \times \frac{3}{4}$

*Estimate:* $1 \times 1 = 1$

$1\frac{1}{4} \times \frac{3}{4} = \frac{5}{4} \times \frac{3}{4}$ *Express $1\frac{1}{4}$ as an improper fraction.*

$= \frac{15}{16}$ *Multiply and then compare with your estimate.*

**B** $2\frac{2}{3} \times 5\frac{1}{2}$

$\frac{8}{3} \times \frac{11}{2}$ *Estimate $3 \times 5 = 15$ and then rewrite the mixed numbers as improper fractions.*

$= \frac{\cancel{8}^{4} \cdot 11}{3 \cdot \cancel{2}_{1}}$ *The GCF of 8 and 2 is 2. Divide both the numerator and denominator by 2 and then multiply.*

$= \frac{44}{3}$ *or* $14\frac{2}{3}$ *Rewrite as a mixed number and compare with your estimate.*

### Try These Together

***Multiply. Write in simplest form.***

**1.** $\frac{4}{5} \times 3\frac{1}{5}$

*HINT: Rewrite the mixed number as an improper fraction and multiply.*

**2.** $1\frac{1}{3} \times 2\frac{3}{8}$

*HINT: Simplify before you multiply.*

## PRACTICE

***Multiply. Write in simplest form.***

**3.** $4\frac{2}{3} \times 1\frac{1}{8}$ **4.** $3\frac{1}{3} \times 4\frac{1}{2}$ **5.** $1\frac{7}{9} \times 2\frac{1}{7}$

**6.** $4\frac{2}{5} \times 1\frac{4}{11}$ **7.** $2\frac{4}{9} \times 2\frac{7}{10}$ **8.** $2\frac{4}{5} \times 9\frac{1}{6}$

**9. Standardized Test Practice** It takes Julie $2\frac{1}{4}$ minutes to run once around a track. How long will it take her to run $8\frac{1}{2}$ laps?

**A** $19\frac{1}{8}$ minutes **B** $19\frac{1}{4}$ minutes **C** $18\frac{7}{8}$ minutes **D** $18\frac{3}{4}$ minutes

Answers: **1.** $2\frac{14}{25}$ **2.** $3\frac{1}{6}$ **3.** $5\frac{1}{4}$ **4.** 15 **5.** $3\frac{17}{21}$ **6.** 6 **7.** $6\frac{3}{5}$ **8.** $25\frac{2}{3}$ **9.** A

NAME ______________________ DATE ____________ PERIOD _____

# 7-4 Dividing Fractions (pages 272–275)

Any two numbers whose product is 1 are called **reciprocals**. For example, $\frac{1}{2}$ and 2 are reciprocals because $\frac{1}{2} \cdot 2 = 1$. You use reciprocals when you divide by fractions.

| Dividing Fractions | To divide by a fraction, multiply by its reciprocal. |
|---|---|

## EXAMPLES

**A** Find the reciprocal of $\frac{2}{3}$.

*Since $\frac{2}{3} \cdot \frac{3}{2} = 1$, the reciprocal of $\frac{2}{3}$ is $\frac{3}{2}$.*

**B** Find $\frac{4}{5} \div \frac{1}{3}$.

$\frac{4}{5} \div \frac{1}{3} = \frac{4}{5} \times \frac{3}{1}$ — *Multiply by the reciprocal of $\frac{1}{3}$.*

$= \frac{12}{5}$ or $2\frac{2}{5}$ — *Multiply the numerators and denominators. Rewrite the improper fraction as a mixed number.*

## Try These Together

**1.** Find the reciprocal of $\frac{2}{7}$.

*HINT: What times $\frac{2}{7}$ equals 1?*

**2.** Find $\frac{7}{8} \div \frac{3}{4}$.

*HINT: Multiply by the reciprocal. Simplify before you multiply.*

## PRACTICE

***Find the reciprocal of each number.***

**3.** $\frac{7}{8}$ **4.** 5 **5.** $\frac{3}{5}$

**6.** $\frac{1}{14}$ **7.** $\frac{1}{7}$ **8.** $\frac{9}{10}$

***Divide. Write in simplest form.***

**9.** $\frac{1}{3} \div \frac{3}{4}$ **10.** $\frac{5}{8} \div \frac{1}{6}$ **11.** $\frac{4}{9} \div \frac{1}{5}$

**12.** $\frac{2}{9} \div \frac{3}{4}$ **13.** $\frac{1}{2} \div \frac{1}{16}$ **14.** $\frac{4}{5} \div \frac{3}{8}$

**15. Standardized Test Practice** After the initial fee of \$2.00, a taxi ride costs \$0.25 per $\frac{1}{5}$ mile. How much would a 4 mile cab ride cost, including the initial fee?

**A** \$5.00 **B** \$3.00 **C** \$20.00 **D** \$7.00

Answers: 1. $\frac{7}{2}$ 2. $1\frac{1}{6}$ 3. $\frac{8}{7}$ 4. $\frac{1}{5}$ 5. $\frac{5}{3}$ 6. 14 7. 7 8. $\frac{10}{9}$ 9. $\frac{4}{9}$ 10. $3\frac{3}{4}$ 11. $2\frac{2}{9}$ 12. $\frac{8}{27}$ 13. 8 14. $2\frac{2}{15}$ 15. D

NAME ______________________ DATE ____________ PERIOD ______

# 7-5 Dividing Mixed Numbers (pages 276–279)

When you divide mixed numbers, first rewrite the mixed numbers as improper fractions. Then divide as you would with a fraction—by multiplying by the reciprocal.

**EXAMPLES**

**A** Find the reciprocal of $4\frac{1}{5}$.

$4\frac{1}{5} = \frac{21}{5}$ *Rewrite as an improper fraction.*

*Since* $\frac{21}{5} \times \frac{5}{21} = 1$, *the reciprocal of* $4\frac{1}{5}$ *is* $\frac{5}{21}$.

**B** Find $2\frac{2}{3} \div 3\frac{1}{2}$.

$2\frac{2}{3} \div 3\frac{1}{2} = \frac{8}{3} \div \frac{7}{2}$ *Rewrite mixed numbers as improper fractions.*

$= \frac{8}{3} \times \frac{2}{7}$ *Multiply by the reciprocal.*

$= \frac{16}{21}$

## Try These Together

**1.** Find the reciprocal of $1\frac{5}{7}$.

*HINT: Rewrite the mixed number as an improper fraction.*

**2.** Find $3\frac{3}{5} \div 8\frac{1}{5}$.

*HINT: Rewrite the mixed numbers as improper fractions. Multiply by the reciprocal.*

**PRACTICE**

***Write each mixed number as an improper fraction. Then write its reciprocal.***

**3.** $7\frac{1}{6}$ **4.** $3\frac{1}{2}$ **5.** $1\frac{7}{8}$ **6.** $2\frac{4}{9}$

**7.** $5\frac{3}{5}$ **8.** $6\frac{1}{8}$ **9.** $2\frac{5}{8}$ **10.** $1\frac{4}{7}$

***Divide. Write in simplest form.***

**11.** $2\frac{2}{5} \div 1\frac{1}{11}$ **12.** $3\frac{1}{6} \div \frac{1}{3}$ **13.** $1\frac{2}{3} \div 4$

**14.** $4\frac{1}{3} \div \frac{6}{7}$ **15.** $\frac{2}{5} \div 1\frac{1}{12}$ **16.** $3\frac{1}{10} \div 2\frac{1}{5}$

**17.** $2\frac{4}{9} \div 1\frac{1}{9}$ **18.** $4\frac{1}{2} \div 2\frac{2}{5}$ **19.** $2\frac{1}{8} \div \frac{1}{2}$

**20. Standardized Test Practice** A sand mosaic requires $\frac{1}{4}$ cup of sand per project. If there are $3\frac{3}{4}$ cups of sand available, how many mosaics can be completed?

**A** 9 **B** 12 **C** 15 **D** 18

Answers: **1.** $\frac{7}{12}$ **2.** $\frac{18}{41}$ **3.** $\frac{43}{6}, \frac{6}{43}$ **4.** $\frac{7}{2}, \frac{2}{7}$ **5.** $\frac{15}{8}, \frac{8}{15}$ **6.** $\frac{22}{9}, \frac{9}{22}$ **7.** $\frac{28}{5}, \frac{5}{28}$ **8.** $\frac{49}{8}, \frac{8}{49}$ **9.** $\frac{21}{8}, \frac{8}{21}$ **10.** $\frac{11}{7}, \frac{7}{11}$ **11.** $2\frac{1}{5}$ **12.** $9\frac{1}{2}$ **13.** $\frac{5}{12}$ **14.** $5\frac{1}{18}$ **15.** $\frac{24}{65}$ **16.** $1\frac{9}{22}$ **17.** $2\frac{1}{5}$ **18.** $1\frac{7}{8}$ **19.** $4\frac{1}{4}$ **20.** C

NAME ______________________ DATE __________ PERIOD _____

# 7-6 Sequences (pages 282–284)

A **sequence** is a list of numbers in a specific order. For example, the numbers 3, 6, 9, 12, 15 are a sequence. In this sequence, notice that 3 is added to each number. The next number in the sequence is 15 + 3, or 18. There are also sequences in which you find the numbers by multiplying by the same number.

## EXAMPLES

***Describe each pattern. Then find the next number in each sequence.***

**A** 13, 18, 23, 28, …

*In this sequence, 5 is added to each number. The next number is 28 + 5, or 33.*

**B** 5, 10, 20, 40, …

*Each number in this sequence is multiplied by 2. The next number is 40 × 2, or 80.*

### Try These Together

***Describe each pattern. Then find the next number in each sequence.***

**1.** 63, 59, 55, 51, …

*HINT: What number is subtracted from each number in the sequence?*

**2.** $2\frac{1}{2}$, 5, $7\frac{1}{2}$, 10, …

*HINT: What number is added to each number in the sequence?*

## PRACTICE

***Describe each pattern. Then find the next two numbers in each sequence.***

**3.** 114, 57, $28\frac{1}{2}$, …

**4.** $\frac{1}{16}$, $\frac{1}{8}$, $\frac{1}{4}$, …

**5.** 14, $16\frac{1}{2}$, 19, …

**6.** 2, 16, 128, …

**7.** $\frac{1}{4}$, $\frac{3}{4}$, $2\frac{1}{4}$, …

**8.** 31, 34, 37, …

***Find the missing number in each sequence.***

**9.** 4, __?__, 36, 108

**10.** 59, __?__, 50, $45\frac{1}{2}$

**11.** $\frac{1}{4}$, $2\frac{1}{2}$, __?__, 250

**12.** $\frac{1}{8}$, $\frac{5}{8}$, __?__, $1\frac{5}{8}$

**13.** 5, 20, 35, __?__

**14.** __?__, 90, 62, 34

**15.** **Standardized Test Practice** Team A is playing Team B in a baseball game. By the end of the fifth inning, how many total outs has each team gotten? (There are 3 outs per inning per team.)

**A** 18 **B** 25 **C** 15 **D** 12

**Answers: 1.** subtract 4; 47 **2.** add $2\frac{1}{2}$; $12\frac{1}{2}$ **3.** multiply by $\frac{1}{2}$; $14\frac{1}{4}$, $7\frac{1}{8}$ **4.** multiply by 2; $\frac{1}{2}$, 1 **5.** add $2\frac{1}{2}$; $21\frac{1}{2}$, 24 **6.** multiply by 8; 1,024, 8,192 **7.** multiply by 3; $6\frac{3}{4}$, $20\frac{1}{4}$ **8.** add 3; 40, 43 **9.** 12 **10.** $54\frac{1}{2}$ **11.** 25 **12.** $1\frac{1}{8}$ **13.** 50 **14.** 118 **15.** C

NAME ______________________ DATE ______________ PERIOD _____

# Chapter 7 Review

## Chef's Secret

Chefs often have to change the amounts of ingredients that they use in their recipes when they change the size of the recipes. Help Chef Ramirez change the amounts shown in the measuring cups below. Shade in the new amounts in the empty measuring cups.

**1.** 

$\times \frac{1}{3} =$

**2.** 

$\div \frac{3}{4} =$

**3.** 

$\div 2\frac{1}{2} =$

Answers are located on p. 106.

# 8-1 Integers (pages 294–298)

An **integer** is any number from this set of the whole numbers and their opposites: {… −3, −2, −1, 0, 1, 2, 3, …}.

| | |
|---|---|
| **Writing and Graphing Integers** | • Integers that are greater than zero are **positive integers.** You can write positive integers with or without a + sign.<br>• Integers that are less than zero are **negative integers**. You write negative integers with a − sign.<br>• Zero is the only integer that is neither positive nor negative.<br>• Each integer has an **opposite** that is the same distance from zero but in the opposite direction on the number line. |
| **Comparing Integers** | • Recall that 7 > 3 means 7 is greater than 3. −7 < 3 means that −7 is less than 3.<br>• To order integers, first graph them on a number line. Then write them in order from left to right, or least to greatest. |

## EXAMPLES

**A** Graph 5 and its opposite on a number line.

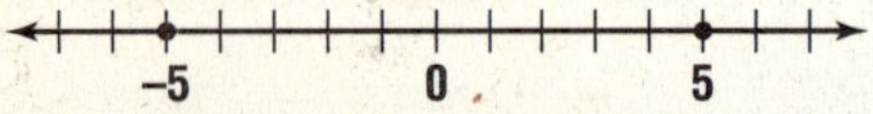


*A number line always has arrows on both ends, with zero and at least one other number marked to show the size of a unit. Make a dot to show the integers you are graphing.*

**B** Which is greater, −7 or −3?

*Think of both of these on a number line. Which integer is to the left? A number to the left is always less than the number to the right.*

*−7 < −3 or −3 > −7*

### Try These Together

**1.** Order from least to greatest: 2, −2, 5, −5, 0.

**2.** Write an integer to represent a debt of $9.

## PRACTICE

***Draw a number line from −10 to 10. Graph each integer on the number line.***

**3.** 2 **4.** −4 **5.** −6 **6.** 5

***Replace each ● with <, >, or = to make a true sentence.***

**7.** −3 ● 5 **8.** 8 ● 2 **9.** −9 ● −9 **10.** −7 ● −12

**11.** Order −5, 6, − 9, and −1 from least to greatest.

**12. Standardized Test Practice** Which integer is the opposite of −25?

**A** −25 **B** 25 **C** 5 **D** −5

**Answers: 1.** −5, −2, 0, 2, 5 **2.** −$9 **3–6.** See Answer Key. **7.** < **8.** > **9.** = **10.** > **11.** −9, −5, −1, 6 **12.** B

NAME ____________________ DATE __________ PERIOD ______

# 8-2 Adding Integers (pages 300–303)

You can use a number line to add integers.

| | |
|---|---|
| **Adding Integers** | To find the sum of $5 + (-7)$, follow these steps.<br>• Start at zero on the number line.<br>• Go 5 in the positive direction (right).<br>• From that point, go 7 in the negative direction (left).<br>• The point where you end $(-2)$ is the sum. |

## EXAMPLES

**A** Is this sum positive, negative, or zero?
$-3 + 5$
*Which integer is farther from zero? +5. The sum will have the same sign as the integer that is farther from zero. The sum of $-3 + 5$ is positive.*

**B** Find the sum of $-4 + (-3)$.
*Start at zero on the number line. Go 4 in the negative direction. From that point, go 3 more in the negative direction. You end at the point $-7$. The sum of $-4 + (-3)$ is $-7$.*

## Try These Together

**1.** Is $8 + (-10)$ positive, negative, or zero?
*HINT: Which integer is farther from zero?*

**2.** Find the sum of $-12 + 13$.
*HINT: Use a number line.*

## PRACTICE

***Tell whether each sum is* positive, negative, *or* zero *without adding.***

**3.** $2 + 4$ **4.** $5 + (-10)$ **5.** $-8 + (-2)$ **6.** $-3 + (-3)$

**7.** $-1 + 5$ **8.** $-4 + (-4)$ **9.** $5 + (-3)$ **10.** $6 + (-6)$

***Add.***

**11.** $-8 + 16$ **12.** $15 + (-5)$ **13.** $4 + (-3)$ **14.** $-7 + 5$

**15.** $3 + (-5)$ **16.** $-2 + (-2)$ **17.** $-6 + 3$ **18.** $8 + (-4)$

**19.** What is 2 plus $-4$ plus $-3$?

**20.** Find the sum of $-14$ and 22.

**21. Football** In a football game, team A was on the 50 yard line. Then they lost 7 yards on the next play. What yard line are they on now?

**22. Standardized Test Practice** What is the sum of $-8$, 4, and $-2$?

**A** $-6$ **B** $-8$ **C** 4 **D** 2

**Answers:** **1.** negative **2.** 1 **3.** positive **4.** negative **5.** negative **6.** negative **7.** positive **8.** negative **9.** positive **10.** zero **11.** 8 **12.** 10 **13.** 1 **14.** $-2$ **15.** $-2$ **16.** $-4$ **17.** $-3$ **18.** 4 **19.** $-5$ **20.** 8 **21.** 43 yard line **22.** A

NAME ______________________ DATE ____________ PERIOD _____

# 8-3 Subtracting Integers (pages 304–307)

You can use counters or a number line to subtract integers.

| | |
|---|---|
| **Subtracting Integers** | To find the difference $4 - (-7)$, follow these steps.<br>• Place 4 positive counters on a mat.<br>• To subtract $-7$, you must remove 7 negative counters. To be able to do this, first add 7 zero pairs to the mat.<br>• Remove 7 negative counters. There are 11 positive counters remaining on the mat.<br>• $4 - (-7) = 11$ |

## EXAMPLES

**A** Find $-3 - (-4)$.

*Begin with 3 negative counters. Add a zero pair to the mat, then remove 4 negative counters. There is 1 positive counter remaining.*

$-3 - (-4) = 1$

**B** Find $6 - 8$.

*Start at zero on the number line and go to 6. From there go 8 in the negative direction (left). You end at $-2$.*

$6 - 8 = -2$

### Try These Together

**1.** Find $6 - (-9)$.

*HINT: Begin with 6 positive counters, then add 9 zero pairs.*

**2.** Find $-9 - 3$.

*HINT: Start at zero and go 9 in the negative direction. From there, go 3 more in the negative direction.*

## PRACTICE

***Subtract. Use counters or a number line.***

**3.** $4 - 2$ **4.** $3 - 5$ **5.** $4 - 7$ **6.** $5 - 1$

**7.** $4 - (-5)$ **8.** $-3 - (-3)$ **9.** $6 - 9$ **10.** $-10 - 5$

**11.** $-7 - (-2)$ **12.** $14 - (-1)$ **13.** $8 - (-3)$ **14.** $-9 - 4$

**15.** Find $-3 + 2 - (-6)$.

**16.** Find the value of $x - y$ if $x = -7$ and $y = 3$.

**17. Landscaping** Charlie is a landscaper. He planted a row of flowers 2 feet back from the street. He then planted a row of bushes 4 feet behind the flowers. What negative integer represents how far back from the street the row of bushes is?

**18.** **Standardized Test Practice** What is the difference $-15 - (-5)$?

**A** $-10$ **B** $-20$ **C** 10 **D** 20

Answers: **1.** 15 **2.** −12 **3.** 2 **4.** −2 **5.** −3 **6.** 4 **7.** 9 **8.** 0 **9.** −3 **10.** −15 **11.** −5 **12.** 15 **13.** 11 **14.** −13 **15.** 5 **16.** −10 **17.** −6 **18.** A

NAME ______________________ DATE ____________ PERIOD _____

# 8-4 Multiplying Integers (pages 310–313)

Remember that multiplication is repeated addition. You can multiply integers by using counters or by using a number line to show repeated addition.

| | |
|---|---|
| **Multiplying Integers** | $4 \times (-3)$ means to put 4 sets of 3 negative counters on a mat. Then count the counters. There are 12 negative counters, so $4 \times (-3) = -12$.<br>$-4 \times (-3)$ means to remove 4 sets of 3 negative counters. To be able to do this, you must first place 4 sets of 3 zero pairs on the mat. Then remove the 4 sets of 3 negative counters. There are 12 positive counters remaining on the mat, so $-4 \times (-3) = 12$. |

## EXAMPLES

**A** Find $-3 \times 5$.

*$-3 \times 5$ means to remove 3 sets of 5 positive counters. Begin with 3 sets of 5 zero pairs on a mat. Then remove the 3 sets of 5 positive counters. There are 15 negative counters remaining, so $-3 \times 5 = -15$.*

**B** Find $2(-11)$.

*You can also use a number line. Begin at zero. Move 11 units to the left, then 11 more units to the left. You end at $-22$, so $2(-11) = -22$.*

## Try These Together

**1.** What is the product of $-4$ and $-8$?

*HINT: Begin with 4 sets of 8 zero pairs. Then remove the 4 sets of 8 negative counters.*

**2.** Find the product of 6 and $-2$.

*HINT: Begin at zero on a number line. Move 2 units to the left 6 times.*

## PRACTICE

***Multiply.***

**3.** $1 \times (-1)$ **4.** $5 \times 4$ **5.** $-3 \times (-3)$ **6.** $-6 \times 2$

**7.** $8 \times (-4)$ **8.** $-3 \times (-7)$ **9.** $5 \times (-3)$ **10.** $-1 \times 9$

**11.** $-6(-4)$ **12.** $-10(-3)$ **13.** $7(-5)$ **14.** $8(-9)$

**15.** Solve $12(-3) = a$.

**16.** What is the product of $-8$ and $-2$?

**17. Time** In winter, the days get shorter until December 21st. If each day is 2 minutes shorter than the day before, how many minutes will be lost in 5 days?

**18.** **Standardized Test Practice** Find the product of $-7$ and $-4$.

**A** 3 **B** $-11$ **C** 28 **D** $-21$

Answers: **1.** 32 **2.** −12 **3.** −1 **4.** 20 **5.** 9 **6.** −12 **7.** −32 **8.** 21 **9.** −15 **10.** −9 **11.** 24 **12.** 30 **13.** −35 **14.** −72 **15.** −36 **16.** 16 **17.** 10 minutes **18.** C

NAME ______________________ DATE ____________ PERIOD ______

# 8-5 Dividing Integers (pages 316–319)

You can model dividing integers with counters or you can use a pattern.

| | |
|---|---|
| **Dividing Integers** | • When you divide two positive integers, or two negative integers, the quotient is positive.<br>• When you divide a negative integer and a positive integer, the quotient is negative. |

## EXAMPLES

**A** Find $-15 \div (-3)$.

*The signs are the same. The quotient is positive.*

$-15 \div (-3) = 5$

**B** Find $6 \div (-2)$.

*The signs are different. The quotient is negative.*

$6 \div (-2) = -3$

### Try These Together

**1.** Find $-12 \div 3$.

*HINT: If you divide 12 negative counters into 3 groups, how many negative counters are in each group?*

**2.** Find $-20 \div (-5)$.

*HINT: Do the two integers have the same sign or different signs?*

## PRACTICE

***Divide.***

**3.** $8 \div 2$ **4.** $6 \div (-3)$ **5.** $-2 \div 1$ **6.** $10 \div 5$

**7.** $14 \div (-7)$ **8.** $-12 \div (-3)$ **9.** $24 \div (-6)$ **10.** $-1 \div (-1)$

**11.** $-16 \div 4$ **12.** $-9 \div (-3)$ **13.** $4 \div 2$ **14.** $-5 \div (-1)$

**15.** Find the value of $-32 \div -16$.

**16.** Divide 42 by $-7$.

**17. Stock Market** Mr. Jimenez lost $320 in 4 days in the stock market. How much money did he lose each day?

**18. Plumbing** The Chens' kitchen faucet has a leak. It drips 3 quarts of water every day. How many quarts of water does it drip in one week?

**19. Standardized Test Practice** What is $-81$ divided by 9?

**A** $-8$ **B** 8 **C** $-9$ **D** 9

Answers: 1. −4 2. 4 3. 4 4. −2 5. −2 6. 2 7. −2 8. 4 9. −4 10. 1 11. −4 12. 3 13. 2 14. 5 15. 2 16. −6 17. $80 18. 21 quarts 19. C

# 8-6 The Coordinate Plane (pages 320–323)

A **coordinate plane** consists of a horizontal line (called the ***x*-axis**) and a vertical line (called the ***y*-axis**) that intersect at the **origin**.

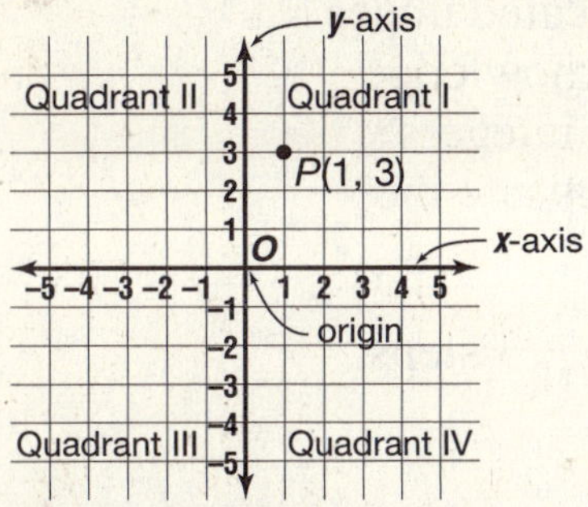


- The *x*-axis and the *y*-axis divide the plane into four **quadrants.**
- You can name point *P* with an **ordered pair** of numbers. The order makes a difference. The pair (1,3) is not the same as (3,1).
- The first number in the pair tells you how far to move to the right or left of the origin. It is called the *x*-coordinate.
- The second number in the pair tells you how far to move up or down from the *x*-axis. It is called the *y*-coordinate.

## EXAMPLES

**A** Give the ordered pair for the point which is 2 units to the right of the origin and 3 units down.

*Show movements to the right and up with positive integers and movements to the left and down with negative integers. This ordered pair is (2, −3).*

**B** What is the ordered pair for the point 4 units to the left of the origin and 5 units up?

*Since you move to the left, the x-coordinate is negative. This ordered pair is (−4, 5).*

### Try These Together

**1.** What are the coordinates of the origin?
*HINT: How much will you move from zero?*

**2.** What is true of all points in Quadrant III?
*HINT: Which ways do you move from the origin to get to a point in Quadrant III?*

## PRACTICE

***Write the ordered pair that names each point.***

**3.** *D* **4.** *A* **5.** *I*

**6.** *C* **7.** *G* **8.** *H*

**9.** *B* **10.** *F* **11.** *E*

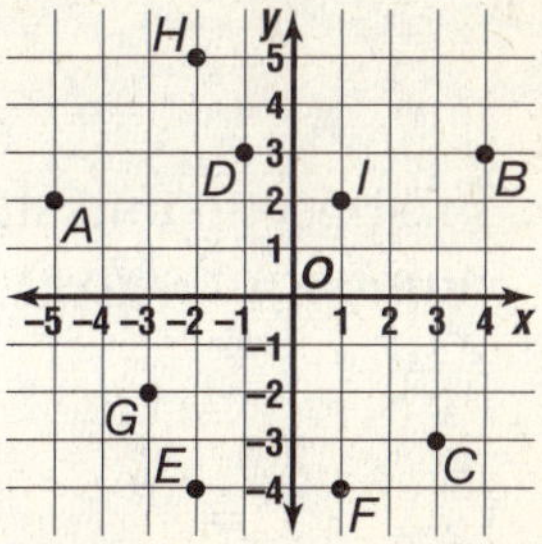


**12. Standardized Test Practice** In which quadrant is Point *J* located?

**A** Quadrant I **B** Quadrant II

**C** Quadrant III **D** Quadrant IV

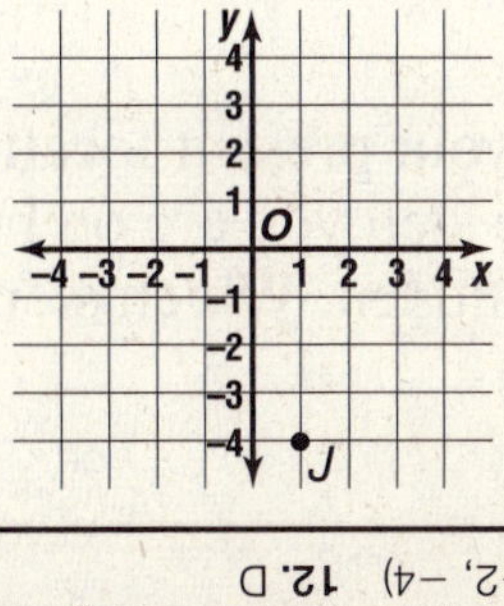


**Answers: 1.** (0, 0) **2.** Both the coordinates are negative. **3.** (−1, 3) **4.** (−5, 2) **5.** (1, 2) **6.** (3, −3) **7.** (−3, −2) **8.** (−2, 5) **9.** (4, 3) **10.** (1, −4) **11.** (−2, −4) **12.** D

# 8 Chapter 8 Review

## Up-and-Down Scavenger Hunt

You've entered a haunted house with your friends. The only way you can get out is to find the key to give to the doorkeeper. The key is located under a board on one of the steps on the staircase. You must use your knowledge of integers to find the step where the key is located. All positive integers indicate the number of steps you go up, and negative integers indicate the number of steps you go down.

1. Starting at the bottom of the staircase, go up 5 steps. Then go $-3$ steps. On which step are you located?

2. From your present location, go to the step that is 3 times the value of your current step. On which step are you now?

3. Subtract $-11$ steps from your location and go to the corresponding step. Where are you now?

4. First go up one step and then divide the step you are on by $-3$ to find the number of steps you take next. On which step did you end up?

5. Add $-8$ steps to your present location and go to the corresponding step. Then multiply the step you are on by 4. The product is the step under which the key is hidden. Which step is it?

Answers are located on p. 106.

NAME ______________________ DATE ____________ PERIOD ______

# 9-1 Properties (pages 333–336)

**Properties** are statements that are true for all values of the variables.

| | | |
|---|---|---|
| **Distributive Property** | To multiply a sum by a number, multiply each addend of the sum by the number outside the parentheses. | $3(5 + 2) = 3 \times 5 + 3 \times 2$<br>$a(b + c) = ab + ac$ |
| **Commutative Property** | The order in which numbers are added or multiplied does not change the sum or product. | $6 + 8 = 8 + 6$<br>$7 \times 4 = 4 \times 7$ |
| **Associative Property** | The way in which numbers are grouped when added or multiplied does not change the sum or product. | $(2 + 5) + 3 = 2 + (5 + 3)$<br>$(6 \times 9) \times 4 = 6 \times (9 \times 4)$ |
| **Additive Identity** | The sum of any number and 0 is the number. | $4 + 0 = 4 \quad a + 0 = a$ |
| **Multiplicative Identity** | The product of any number and 1 is the number. | $5 \times 1 = 5 \quad 1 \times n = n$ |

## EXAMPLES

**A** Find $5 \times 12$ mentally using the Distributive Property.

$5 \times 12 = 5(10 + 2)$ *Use 10 + 2 for 12.*
$= 5(10) + 5(2)$
$= 50 + 10 = 60$

**B** Find $8 + 11 + 2 + 9$ mentally.

$8 + 11 + 2 + 9$
$= 8 + 2 + 11 + 9$ *Commutative Property*
$= (8 + 2) + (11 + 9)$ *Associative Property*
$= 10 + 20 = 30$ *Add mentally.*

### Try These Together

***Find each product mentally. Use the Distributive Property. Then evaluate.***

**1.** $9 \times 17$ **2.** $16 \times 4$

## PRACTICE

***Rewrite each expression using the Distributive Property. Then evaluate.***

**3.** $7(60 + 8)$ **4.** $8(50 + 1)$ **5.** $52 \times 50 + 52 \times 6$

***Identify the property shown by each equation.***

**6.** $9 + 0 = 9$ **7.** $65 \times 1 = 65$ **8.** $4 + (7 + 5) = (4 + 7) + 5$

***Find each sum or product mentally.***

**9.** $5 \times 4 \times 8$ **10.** $15 + 14 + 16$ **11.** $2 \times 9 \times 50$

**12. Standardized Test Practice** Find $1.8 \times 5$ mentally.

**A** 0.9 **B** 5.4 **C** 9 **D** 54

**Answers: 1.** 153 **2.** 64 **3.** $7 \times 60 + 7 \times 8$; 476 **4.** $8 \times 50 + 8 \times 1$; 408 **5.** $52(50 + 6)$; 2,912 **6.** Identity(+) **7.** Identity(×) **8.** Assoc.(+) **9.** 160 **10.** 45 **11.** 900 **12.** C

NAME ______________________ DATE ____________ PERIOD ______

# 9-2 Solving Addition Equations (pages 339–342)

You can use models to solve addition equations. You can then use the same pattern as you solve addition equations with paper and pencil.

| | |
|---|---|
| **Solving Addition Equations** | To solve an equation, you get the variable by itself on one side of the equation.<br>To solve an addition equation<br>• Circle the variable you will get by itself on one side of the equation.<br>• Ask yourself, "What do I need to do to undo what has been done to this variable?"<br>• Then do the same thing to each side of the equation. Your variable will then be by itself on one side of the equation, and your numbers will be on the other side of the equation. |

## EXAMPLES

**A** Solve $8 + y = 10$.

| | |
|---|---|
| $8 + y = 10$ | *To get y alone, you must undo adding 8.* |
| $8 + y = 10$ | *Subtract to undo adding 8.* |
| $\underline{-8} \quad \underline{-8}$ | *Subtract 8 from each side.* |
| $y = 2$ | |
| $8 + 2 = 10$ ✓ | *Check by replacing y with 2.* |

**B** Find the value of $n$ if $n + (-2) = 7$.

| | |
|---|---|
| $n + (-2) = 7$ | *To get n alone, you must undo adding (−2).* |
| $n + (-2) = 7$ | *+2 is the opposite of (−2).* |
| $\underline{+2} \quad \underline{+2}$ | *Do the same thing to each side.* |
| $n = 9$ | |
| $9 + (-2) = 7$ ✓ | *Check by replacing n with 9.* |

### Try These Together

**1.** Solve $-3 = b + 4$.
*HINT: You can either subtract 4 or add (−4) to each side of the equation.*

**2.** Solve $t + 5 = -14$.
*HINT: Subtract 5 from each side of the equation.*

## PRACTICE

***Solve each equation. Use models if necessary. Check your solution.***

**3.** $x + 7 = 11$
**4.** $y + 2 = 6$
**5.** $10 + m = 13$
**6.** $2 + n = 11$
**7.** $r + (-1) = 4$
**8.** $16 + t = 26$
**9.** $12 + w = -2$
**10.** $4 + z = 9$
**11.** $d + (-5) = -8$

**12.** Find the value of $a$ if $a + 13 = 26$.

**13.** What is the value of $b$ if $9 + b = -1$?

**14.** **Standardized Test Practice** Find the value of $x$ if $x + 10 = 95$.

**A** 25 **B** 85 **C** 95 **D** 75

Answers: 1. −7 2. −19 3. 4 4. 4 5. 3 6. 9 7. 5 8. 10 9. −14 10. 5 11. −3 12. 13 13. −10 14. B

NAME ______________________ DATE ____________ PERIOD _____

# 9-3 Solving Subtraction Equations (pages 344–347)

You can use models to solve subtraction equations. You can also rewrite a subtraction equation as an addition equation and solve with paper and pencil.

| **Solving Subtraction Equations** | To solve an equation, you get the variable by itself on one side of the equation.<br>To solve a subtraction equation<br>• Ask yourself, "What do I need to do to undo what has been done to this variable?"<br>• Then do the same thing to each side of the equation. |
|---|---|

## EXAMPLES

**A** Solve $y - 7 = 12$.

$y - 7 = 12$

$\underline{+7} \quad \underline{+7}$ *Add 7 to each side.*

$y = 19$

$19 - 7 = 12$ ✓ *Check by replacing y with 19.*

**B** Find the value of $n$ if $n - (-2) = 8$.

$n - (-2) = 8$ *To get n alone, you must undo subtracting (−2).*

$n + 2 = 8$ *Subtracting (−2) is the same as adding 2. The opposite of adding 2 is subtracting 2.*

$\underline{-2} \quad \underline{-2}$ *Do the same thing to each side.*

$n = 6$

$6 - (-2) = 8$ ✓ *Check by replacing n with 6.*

### Try These Together

**1.** Solve $x - 4 = -3$.
*HINT: Add 4 to each side.*

**2.** Solve $p - (-7) = -20$.
*HINT: Rewrite as $p + 7 = -20$.*

## PRACTICE

***Solve each equation. Use models if necessary. Check your solution.***

**3.** $h - 5 = 2$ **4.** $g - 8 = 1$ **5.** $-3 = j - 5$

**6.** $k - (-4) = 10$ **7.** $n - (-6) = 12$ **8.** $r - (-1) = 6$

**9.** $t - 7 = 2$ **10.** $s - 16 = 5$ **11.** $d - 8 = -2$

**12.** $f - 10 = 5$ **13.** $w - 4 = -4$ **14.** $x - 9 = 3$

**15.** Find the value of $z$ if $z - 3 = -2$.

**16.** If $q - (-1) = 4$, what is the value of $q$?

**17.** **Standardized Test Practice** Martina spent $1 on a snack after school and had $4 left. How much money did she have before she bought the snack?

**A** $6 **B** $4 **C** $3 **D** $5

Answers: 1. 1 2. −27 3. 7 4. 9 5. 2 6. 6 7. 6 8. 5 9. 9 10. 21 11. 6 12. 15 13. 0 14. 12 15. 1 16. 3 17. D

# 9-4 Solving Multiplication Equations (pages 350–353)

You can use models to solve multiplication equations. You can also solve an equation with paper and pencil by undoing what has been done.

| | |
|---|---|
| **Solving Multiplication Equations** | • You need to get the variable by itself on one side of the equation by undoing what has been done to the variable.<br>• Ask yourself, "What do I need to do to undo what has been done to this variable?"<br>• Divide to undo multiplication.<br>• Do the same to each side of the equation. |

## EXAMPLES

**A** Solve $8y = 24$.

$8y = 24$ *To get y alone, you must undo multiplying by 8.*

$8y = 24$ *Divide to undo the multiplication.*

$\frac{8y}{8} = \frac{24}{8}$ *Divide each side by 8.*

$y = 3$

$8(3) = 24$ ✓ *Check by replacing y with 3.*

**B** Find the value of $n$ if $18 = -3n$.

$18 = -3n$ *To get n alone, you must undo multiplying by −3.*

$\frac{18}{-3} = \frac{-3n}{-3}$ *Divide each side by −3.*

$-6 = n$

$18 = -3(-6)$ ✓ *Check by replacing n with −6.*

### Try These Together

**1.** Solve $2.7p = -10.8$.
*HINT: Divide each side by 2.7.*

**2.** Solve $4q = 36$.
*HINT: Divide each side by 4.*

## PRACTICE

***Solve each equation. Use models if necessary.***

**3.** $3b = 9$ **4.** $2g = -10$ **5.** $16 = 2x$ **6.** $-5q = 25$

**7.** $54 = 6r$ **8.** $15 = 1p$ **9.** $-24 = 8k$ **10.** $10t = 40$

**11.** $-12 = 4a$ **12.** $7m = 63$ **13** $48 = -6d$ **14.** $9c = -45$

**15.** **Standardized Test Practice** Jalisa has to take 3 teaspoons of medicine for her cold every day until the medicine is gone. If there are 33 teaspoons of medicine in the bottle, how many days will she have to take medicine?

**A** 11 **B** 9 **C** 10 **D** 12

**Answers:** **1.** −4 **2.** 9 **3.** 3 **4.** −5 **5.** 8 **6.** −5 **7.** 9 **8.** 15 **9.** −3 **10.** 4 **11.** −3 **12.** 9 **13.** −8 **14.** −5 **15.** A

NAME ______________________ DATE ____________ PERIOD _____

# 9-5 Solving Two-Step Equations (pages 355–357)

A two-step equation involves two different operations such as addition and multiplication. To solve a two-step equation, you work backward, reversing the order of operations.

| Solving Two-Step Equations | To get the variable alone on one side of the equation<br>• First, undo the number that is added or subtracted.<br>• Second, undo the number that multiplies or divides the variable. |
|---|---|

## EXAMPLES

**A** Solve $3x + 7 = -5$.

| | |
|---|---|
| $3x + 7 = -5$ | *To get x alone, undo adding 7 first.* |
| $3x = -12$ | *Subtract 7 from each side.* |
| $3x = -12$ | *Second, undo multiplying by 3.* |
| $\frac{3x}{3} = -\frac{12}{3}$ | *Divide each side by 3.* |
| $x = -4$ | |
| $3(-4) + 7 = -5$ ✓ | *Check by replacing x with −4.* |

**B** Solve $4 - 5p = 14$.

| | |
|---|---|
| $4 - 5p = 14$ | *To get p alone, undo adding 4 first.* |
| $-5p = 10$ | *Subtract 4 from each side.* |
| $-5p = 10$ | *Second, undo multiplying by −5.* |
| $\frac{-5p}{-5} = \frac{10}{-5}$ | *Divide each side by −5.* |
| $p = -2$ | |
| $4 - 5(-2) = 14$ ✓ | *Check by replacing p with −2.* |

### Try These Together

**1.** Solve $3q - 4 = 8$.

*HINT: Add 4 to each side and then divide by 3.*

**2.** Solve $7 = 3y + 1$.

*HINT: First subtract 1 from each side and then divide each side by 3.*

## PRACTICE

***Solve each equation.***

**3.** $2x + 4 = 8$ **4.** $10y + 5 = 45$ **5.** $4z + 2 = 14$ **6.** $5k + 10 = 50$

**7.** $6t - 9 = 9$ **8.** $5m + 10 = 70$ **9.** $8s - 4 = 28$ **10.** $9h - 5 = 40$

**11.** $15 = -3p + 9$ **12.** $14 = -5q - 1$ **13.** $26 = 3j + 2$ **14.** $40 = 2d + 20$

**15.** Five more than twice a number is 37. Find the number.

**16.** Eight less than three times a number is nineteen. What is the number?

**17.** **Standardized Test Practice** Devin spent $34 at the music store. He bought two CDs for the same price each and a case for $10. How much did each CD cost?

**A** $15 **B** $5 **C** $12 **D** $17

Answers: 1. 4 2. 2 3. 2 4. 4 5. 3 6. 8 7. 3 8. 12 9. 4 10. 5 11. −2 12. −3 13. 8 14. 10 15. 16 16. 9 17. C

NAME ______________________________ DATE ______________ PERIOD _____

# 9-6 Functions (pages 362–365)

When you say "$y$ is a **function** of $x$," this means that the value of $y$ depends on the value of $x$. If you know the input value for $x$ and the function rule, you can find the output value for $y$. A **function table** shows you the input ($x$) and output ($y$) values for a certain function rule.

| | |
|---|---|
| **Making Function Tables and Finding Function Rules** | • To find the output values for a function table, substitute the input values for the variable in the function rule.<br>• To find the function rule when you have the function table, study the relationship between each input and output. |

## EXAMPLES

**A** Complete the function table.

| input ($x$) | output ($x + 2$) |
|---|---|
| −1 | |
| 0 | |
| 2 | |

*$-1 + 2 = \mathbf{1}$*

*$0 + 2 = \mathbf{2}$*

*$2 + 2 = \mathbf{4}$*

**B** Find the rule for the function table.

| input ($x$) | output (?) |
|---|---|
| 1 | 2 |
| 2 | 5 |
| 3 | 8 |

*Notice that the output is 1 less than three times x.*

*The rule is* $\mathbf{3x - 1}$.

## Try These Together

**1.** If the input values are 3, 5, and 6, and the corresponding output values are 7, 11, and 13, what is the function rule?
*HINT: Notice that 7 is 1 more than twice 3.*

**2.** If the function rule is $5x + 2$, what is the output for an input of 0?
*HINT: Substitute 0 for x in the rule and simplify.*

## PRACTICE

***Complete each function table.***

**3.**

| input ($x$) | output ($x - 2$) |
|---|---|
| 2 | |
| 4 | |
| 8 | |

**4.**

| input ($x$) | output ($x + 3$) |
|---|---|
| 1 | |
| 3 | |
| 5 | |

**5.** What is the output for an input of 7 if the function rule is $4x$?

**6.** If the output is 4 and the function rule is $x + 3$, what is the input?

**7.** **Standardized Test Practice** If the function rule is $3x - 4$, what is the output for an input of 3?

**A** 12 **B** 9 **C** 4 **D** 5

Answers: **1.** $2x + 1$ **2.** 2 **3.** 0, 2, 6 **4.** 4, 6, 8 **5.** 28 **6.** 1 **7.** D

NAME ______________________ DATE ____________ PERIOD ____

# 9-7 Graphing Functions (pages 366–369)

You can graph a function rule or equation on a coordinate plane.

| | |
|---|---|
| **Graphing Functions** | When you have a function table, graph the function with these steps.<br>• Write ordered pairs (input, output) from the function table.<br>• Graph each ordered pair on the coordinate plane.<br>• Join the graphed points with a line.<br>When you have a function rule, make a function table for 3 or 4 input values and then graph that table with the steps above. |

## EXAMPLE

Graph $y = 2x + 1$.

| input ($x$) | function rule ($2x + 1$) | output ($y$) | ordered pairs ($x, y$) |
|---|---|---|---|
| 0 | 2(0) + 1 | 1 | (0, 1) |
| 1 | 2(1) + 1 | 3 | (1, 3) |
| 2 | 2(2) + 1 | 5 | (2, 5) |

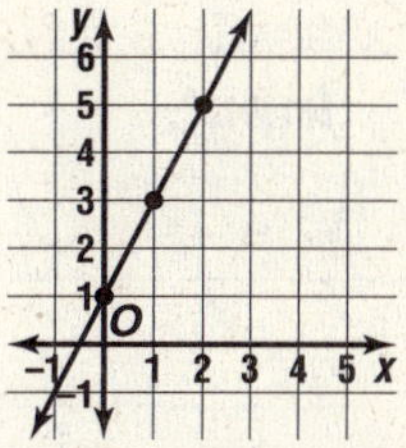


## PRACTICE

***Graph the functions represented by each function table.***

**1.**

| input | output |
|---|---|
| 1 | −1 |
| 3 | 1 |
| 5 | 3 |

**2.**

| input | output |
|---|---|
| −4 | −1 |
| 0 | 3 |
| 4 | 7 |

***Complete each function table. Then graph the function.***

**3.**

| $x$ | $x - 1$ |
|---|---|
| 2 | |
| 4 | |
| 6 | |

**4.**

| $x$ | $x + 4$ |
|---|---|
| −1 | |
| −2 | |
| −3 | |

**5. Fitness** Jakira is training for a triathlon. She runs 3 miles every day. What is the function rule that you could use to determine how far Jakira runs if the input is the number of days?

**6. Standardized Test Practice** What is $y$ (the output) for the function rule $4x$ if $x = 10$?

**A** 6 **B** 40 **C** 80 **D** 4

**Answers:** **1–2.** See Answer Key. **3–4.** See Answer Key for graphs. **3.** 1, 3, 5 **4.** 3, 2, 1 **5.** $3n$ **6.** B

NAME ______________________ DATE ____________ PERIOD ______

# 9 Chapter 9 Review

## Function Flash

You and your parent can use index cards or slips of paper to help you study functions. You can put a function rule and an input value on the front and the output value on the back. Fill in the table below to show what pieces of information you might put on various cards.

| Front | Back |
|---|---|
| $x + 2$<br>5 | 7 |

| | Rule | Input | Output |
|---|---|---|---|
| 1. | $x - 4$ | 2 | |
| 2. | $3x$ | | 9 |
| 3. | $x + 3$ | | 5.3 |
| 4. | $2x - 1$ | $1\frac{1}{2}$ | |
| 5. | $2x + 1$ | | 3 |

6. You can also make cards with input and output values on the front and the function rule on the back. What rule would go on the back of the card shown?

| Input | Output |
|---|---|
| 0 | 4 |
| 2 | 6 |
| 4 | 8 |

Answers are located on p. 106.

NAME ______________________ DATE ____________ PERIOD ____

# 10-1 Ratios (pages 380–383)

You can compare two quantities by using a **ratio**. A common way to express a ratio is as a fraction in simplest form. If the two quantities you are comparing have different units of measure, this kind of ratio is called a **rate**. A rate is in the form of a **unit rate** when the denominator is 1.

| | |
|---|---|
| **Writing a Rate and a Unit Rate** | A rate is a ratio of two measurements that have different units. To write a ratio as a unit rate, divide the numerator and denominator by the same number to rewrite the ratio as a fraction with a denominator of 1. |

## EXAMPLES

**A** Write the ratio 5 sixth-graders out of 15 students in three different ways. Express this ratio as a fraction in simplest form.

*As a fraction* $\frac{5}{15}$

*As a ratio* 5:15

*In words* 5 to 15

*Another way is in the problem: 5 out of 15.*

$\frac{5}{15}$ *in simplest form is* $\frac{1}{3}$.

**B** Express the ratio 15 pencils for $5 as a unit rate. How many pencils can you buy for $1?

*Write the ratio as a fraction.* $\frac{15 \text{ pencils}}{\$5}$

*To rewrite the fraction with a denominator of 1, divide numerator and denominator by 5.*

$\frac{15 \text{ pencils}}{\$5} = \frac{\frac{15}{5} \text{ pencils}}{\$1}$ *or 3 pencils for $1*

## Try These Together

1. Write the ratio 7 sodas out of 20 are sugar free in three different ways.
*HINT: Write the numbers in the same order as they appear in the problem.*

2. Express the ratio $14.50 for 5 rides as a rate. What is the cost for one ride?
*HINT: Divide numerator and denominator by 5.*

## PRACTICE

***Write each ratio as a fraction in simplest form.***

3. 4 out of 16 papers are typed
4. 5 out of 10 horses are white
5. 7 blue bicycles out of 21 bicycles
6. 4 watermelons out of 10 melons

***Write each ratio as a unit rate.***

7. $1.50 for 3 bottles of juice
8. 5 bracelets for $25.00

9. **Standardized Test Practice** If milk costs $5.50 for 2 gallons, how much does it cost per gallon?

**A** $11.00 **B** $10.50 **C** $2.75 **D** $3.50

**Answers: 1.** 7:20, 7 to 20; $\frac{7}{20}$ **2.** $\frac{\$14.50}{5 \text{ rides}}$ or $2.90 per ride **3.** $\frac{1}{4}$ **4.** $\frac{1}{2}$ **5.** $\frac{1}{3}$ **6.** $\frac{2}{5}$ **7.** $0.50 per bottle of juice **8.** $5.00 per bracelet **9.** C

NAME ______________________ DATE ____________ PERIOD _____

# 10-2 Solving Proportions (pages 386–389)

A **proportion** is an equation that shows that two ratios are equivalent. The general form of a proportion is $\frac{a}{b} = \frac{c}{d}$, where neither $b$ nor $d$ is equal to zero. The **cross products** of a proportion are $ad$ and $bc$.

| | |
|---|---|
| **Property of Proportions** | The cross products of a proportion are equal. If $\frac{a}{b} = \frac{c}{d}$, then $ad = bc$. |

## EXAMPLES

**A** Use cross products to find out whether this pair of ratios forms a proportion.

$\frac{3}{4}, \frac{9}{12}$

*Does $\frac{3}{4} = \frac{9}{12}$? Are the cross products equal?*

*Does $3 \times 12 = 4 \times 9$? Yes, because $36 = 36$.*

*$\frac{3}{4} = \frac{9}{12}$ is a proportion because the cross products are equal.*

**B** Solve the proportion $\frac{2}{7} = \frac{y}{21}$ for $y$.

*Write the cross products.*

$2 \times 21 = 7 \times y$

$42 = 7y$

$\frac{42}{7} = \frac{7y}{7}$ *Divide each side of the equation by 7.*

$6 = y$

*The solution is 6.*

## Try These Together

1. Use cross products to determine whether this pair of ratios forms a proportion. $\frac{0.5}{2}, \frac{0.4}{1.6}$

2. Solve the proportion $\frac{3}{p} = \frac{4}{20}$.
*HINT: Set the cross products equal to each other and solve for p.*

## PRACTICE

***Determine whether each pair of ratios forms a proportion.***

3. $\frac{1}{2}, \frac{5}{10}$

4. $\frac{4}{8}, \frac{2}{4}$

5. $\frac{4}{5}, \frac{1}{8}$

6. $\frac{8}{13}, \frac{2}{5}$

***Solve each proportion.***

7. $\frac{3}{6} = \frac{x}{2}$

8. $\frac{4}{w} = \frac{2}{11}$

9. $\frac{9}{12} = \frac{6}{z}$

10. $\frac{d}{25} = \frac{2}{10}$

11. **Standardized Test Practice** The home economics class is making a casserole. They need 3 eggs for 1 casserole. How many eggs do they need for 4 casseroles?

**A** 9 **B** 12 **C** 15 **D** 10

**Answers:** 1. yes 2. 15 3. yes 4. yes 5. no 6. no 7. 1 8. 22 9. 8 10. 5 11. B

NAME ______________________ DATE ____________ PERIOD _____

# 10-3 Scale Drawings and Models (pages 391–393)

A **scale drawing** is exactly the same shape as an object, but the drawing may be larger or smaller than the real object.

| | |
|---|---|
| **Reading a Scale Drawing** | The scale written on the drawing or model gives the ratio that compares the measurements on the drawing to the actual measurements of the object. Use the scale of the drawing for one of the ratios and the known and unknown measurements for the other ratio. Write a proportion and solve it for the unknown measurement. |

## EXAMPLES

**A** A model car has a scale of 1:16. A window on the model measures $\frac{1}{32}$ of a meter. What will this same window measure on the real car?

$\frac{1}{16} = \frac{\frac{1}{32}\text{ meter}}{\ell}$   *Write a proportion.*

$\ell = 16\left(\frac{1}{32}\text{ meter}\right)$   *Find cross products.*

$\ell = \frac{1}{2}\text{ meter}$   *Solve.*

*The actual window measures $\frac{1}{2}$ meter.*

**B** The doorway of an actual house measures 3 ft wide. How wide will the doorway in a model house be if the scale is 1 ft = 2 in.?

$\frac{1\text{ ft}}{2\text{ in.}} = \frac{3\text{ ft}}{w}$ *so 1 x w = 6 or w = 6.*

*The model doorway will be 6 inches wide.*

### Try These Together

**1.** The scale of a map is 1 inch = 25 miles. The distance on the map between two cities is 7 inches. How many miles apart are they?

*HINT: Write a proportion using $\frac{1}{25}$ as one ratio.*

**2.** A line on a scale drawing of a building measures 15 inches. The same length on the actual building is 5 yards. What is the scale of the drawing in simplest form?

*HINT: One ratio is $\frac{15}{5}$ and the other is $\frac{x\text{ inches}}{1\text{ yard}}$.*

## PRACTICE

**3. Transportation** The oldest monorail system in the world is in Wuppertal, Germany. Its track is 8.5 miles long. If you wanted to build a model of the track that has a scale of 1 inch = 0.5 miles, how long would the model track be?

**4.** **Standardized Test Practice** Mavis and Reese want to rearrange the furniture in their living room. Before they move the furniture, they make a model. The scale for the model is 1 inch = 2 feet. If their sofa is actually 6 feet long, how long is the model of the sofa?

**A** 3 inches **B** 4 inches **C** 3 feet **D** 4 feet

**Answers: 1.** 175 miles **2.** 3 inches to 1 yard **3.** 17 inches **4.** A

NAME ______________________ DATE ____________ PERIOD _____

# 10-4 Modeling Percents (pages 395–397)

A **percent** is a ratio that compares a number to 100. Since percent means *out of one hundred,* you can use a 10 × 10 grid to model percents.

## EXAMPLES

**A** Model 20%

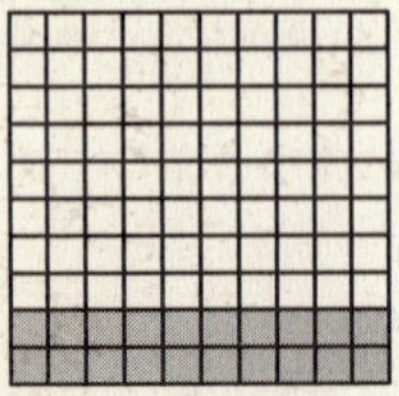

20%

*20% means 20 out of 100. So, shade 20 of the 100 squares.*

**B** Model 36%

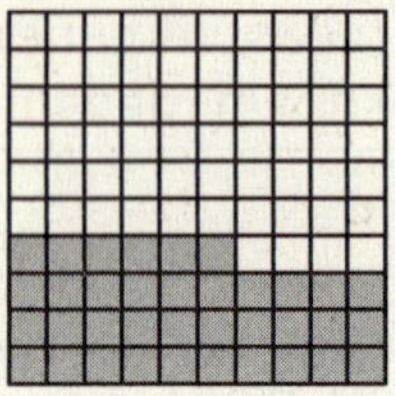

36%

*36% means 36 out of 100. So, shade 36 of the 100 squares.*

## PRACTICE

***Model each percent.***

**1.** 8% **2.** 45% **3.** 17% **4.** 63% **5.** 55% **6.** 90%

***Identify each percent modeled.***

**7.**

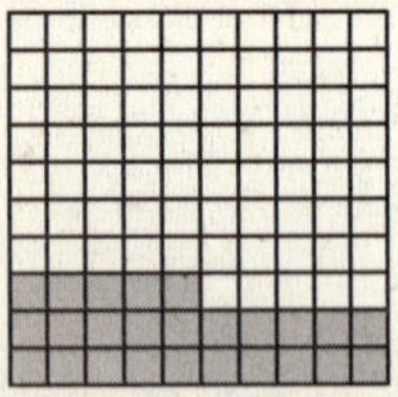

**8.**

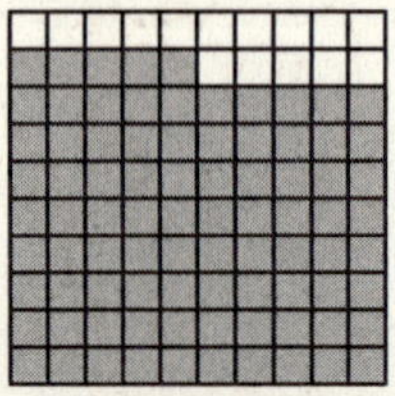

**9.**

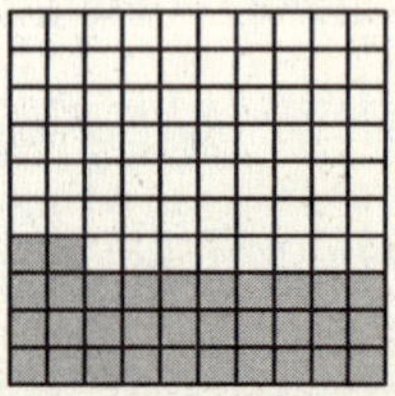

**10.**

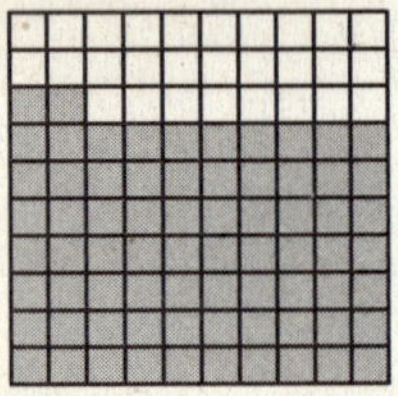

**11.**

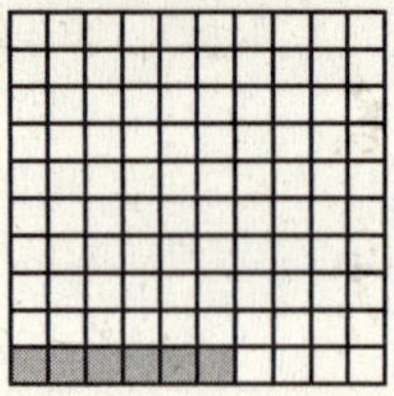

**12.**

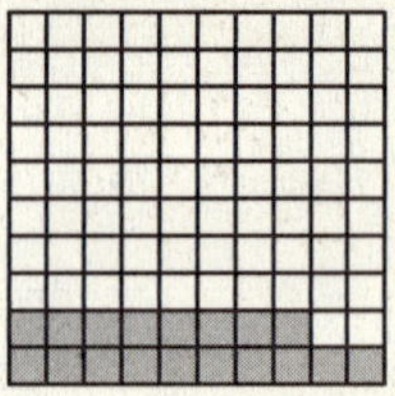

**13.** At the school cafeteria, 65% of the students drink soda. Make a model to show 65%.

**14.** Use a model to show which is smaller, 83% or 77%.

**15.** **Standardized Test Practice** In the eighth grade class, 16% exercise 3 to 4 hours per week, 23% exercise 2 to 3 hours per week, 25% exercise 1 to 2 hours per week, 36% exercise 0 to 1 hours per week. What amount of time spent exercising has the least percentage of students?

**A** 3 to 4 hours **B** 2 to 3 hours **C** 1 to 2 hours **D** 0 to 1 hours

**Answers: 1–6** See Answer Key **7.** 25% **8.** 85% **9.** 32% **10.** 72% **11.** 6% **12.** 18% **13–14.** See Answer Key. **14.** 77% is smaller. **15.** A

NAME ______________________ DATE ____________ PERIOD _____

# 10-5 Percents and Fractions (pages 400–403)

A **percent** is a ratio that compares a number to 100.

| | |
|---|---|
| **Writing a Percent as a Fraction** | To write a percent as a fraction, follow these steps.<br>• Write the percent as a fraction with a denominator of 100.<br>• Simplify the fraction. |
| **Writing a Fraction as a Percent** | To write a fraction as a percent, follow these steps.<br>• Set up a proportion with the fraction as one ratio and $\frac{x}{100}$ as the other.<br>• Find the cross products and divide to solve for $x$. The fraction is equal to $x$ percent. |

## EXAMPLES

**A** Write 75% as a fraction in simplest form.

*75% is $\frac{75}{100}$.*

$75\% = \frac{75}{100}$

$75\% = \frac{3}{4}$ *Divide numerator and denominator by the common factor of 25.*

**B** Write $\frac{14}{25}$ as a percent.

$\frac{14}{25} = \frac{x}{100}$ *Write a proportion.*

$1{,}400 = 25x$ *Find the cross products.*

$\frac{1{,}400}{25} = x$ *Divide to solve for x.*

$56 = x$, so $\frac{14}{25} = 56\%$

## Try These Together

**1.** Write $\frac{13}{20}$ as a percent.

**2.** Write 120% as a fraction in simplest form.

## PRACTICE

***Write each percent as a fraction in simplest form.***

**3.** 25% **4.** 10% **5.** 30% **6.** 45%

**7.** 60% **8.** 95% **9.** 16% **10.** 58%

***Write each fraction as a percent.***

**11.** $\frac{1}{2}$ **12.** $\frac{8}{5}$ **13.** $\frac{3}{4}$ **14.** $\frac{44}{100}$

**15.** $\frac{12}{20}$ **16.** $\frac{36}{40}$ **17.** $\frac{8}{40}$ **18.** $\frac{7}{5}$

**19.** **Standardized Test Practice** Write 24% as a fraction in simplest form.

**A** $\frac{18}{75}$ **B** $\frac{12}{50}$ **C** $\frac{24}{100}$ **D** $\frac{6}{25}$

Answers: 1. 65% 2. $1\frac{1}{5}$ 3. $\frac{1}{4}$ 4. $\frac{1}{10}$ 5. $\frac{3}{10}$ 6. $\frac{9}{20}$ 7. $\frac{3}{5}$ 8. $\frac{19}{20}$ 9. $\frac{4}{25}$ 10. $\frac{29}{50}$ 11. 50% 12. 160 % 13. 75% 14. 44% 15. 60% 16. 90% 17. 20% 18. 140% 19. D

# 10-6 Percents and Decimals (pages 404–406)

You have seen that percents can be written as fractions. Percents can also be written as decimals, and decimals can be written as percents.

| | |
|---|---|
| **Writing a Percent as a Decimal** | To write a percent as a decimal, follow these steps.<br>• Rewrite the percent as a fraction with a denominator of 100.<br>• Write the fraction as a decimal. |
| **Writing a Decimal as a Percent** | To write a decimal as a percent, follow these steps.<br>• Rewrite the decimal as a fraction with a denominator of 100.<br>• Write the fraction as a percent. |

## EXAMPLES

**A** Write 56% as a decimal.

$56\% = \frac{56}{100}$ *which is 0.56*

**B** Write 0.84 as a percent.

$0.84 = \frac{84}{100}$ *which is 84%*

**C** Write 0.35% as a decimal.

$0.35\% = \frac{0.35}{100}$ *Multiply by* $\frac{100}{100}$ *to get rid of the decimal in the numerator.*

$= \frac{35}{10{,}000}$ *which is 0.0035*

**D** Write 0.103 as a percent.

$0.103 = \frac{103}{1{,}000}$ *Divide numerator and denominator by 10.*

$= \frac{10.3}{100}$ *which is 10.3%*

### Try These Together

**1.** Write 0.4% as a decimal.

**2.** Write 0.09 as a percent.

## PRACTICE

***Write each percent as a decimal.***

**3.** 27% **4.** 18% **5.** 46% **6.** 55%

**7.** 72% **8.** 91% **9.** 11% **10.** 34.5%

***Write each decimal as a percent.***

**11.** 0.14 **12.** 0.87 **13.** 0.25 **14.** 0.61

**15.** 0.59 **16.** 0.12 **17.** 0.73 **18.** 0.063

**19.** **Standardized Test Practice** In a taste test at a grocery store, people were given a chip with salsa on it and asked if they would buy the salsa. Of those who answered, 67% said "yes." Express this percent as a decimal.

**A** 0.22 **B** 0.67 **C** 0.34 **D** 0.50

**Answers: 1.** 0.004 **2.** 9% **3.** 0.27 **4.** 0.18 **5.** 0.46 **6.** 0.55 **7.** 0.72 **8.** 0.91 **9.** 0.11 **10.** 0.345 **11.** 14% **12.** 87% **13.** 25% **14.** 61% **15.** 59% **16.** 12% **17.** 73% **18.** 6.3% **19.** B

NAME ______________________ DATE ____________ PERIOD _____

# 10-7 Percent of a Number (pages 409–412)

To find the percent of a number, you can change the percent to a fraction or to a decimal, and then multiply by the number. You can also use a calculator.

| | |
|---|---|
| **Finding the Percent of a Number** | • Method 1: Change the percent to a fraction and multiply.<br>• Method 2: Change the percent to a decimal and multiply. |

## EXAMPLES

**A** Find 25% of 56.

*25% = $\frac{1}{4}$*

*$\frac{1}{4} \times 56 = 14$*

*25% of 56 is 14.*

**B** Find 103% of 60.

*103% = $\frac{103}{100}$ which is 1.03*

*$1.03 \times 60 = 61.8$*

*103% of 60 is 61.8.*

*Notice that when you take a percent greater than 100 of a number, the answer is greater than the number.*

### Try These Together

**1.** Find 0.5% of 30.

*HINT: Rewrite the percent as $\frac{0.5}{100}$ and then as $\frac{5}{1,000}$ or 0.005. Then multiply.*

**2.** Find 7% of 40.

*HINT: Rewrite 7% as $\frac{7}{100}$ or 0.07.*

## PRACTICE

***Find the percent of each number.***

**3.** 25% of 20 **4.** 40% of 65 **5.** 35% of 80 **6.** 60% of 35

**7.** 80% of 120 **8.** 75% of 64 **9.** 10% of 70 **10.** 20% of 45

**11.** 33% of 300 **12.** 20% of 120 **13.** 50% of 64 **14.** 90% of 60

**15.** What is 90% of 70?

**16.** Find 80% of 80.

**17. Games** 75% of the games sold at a game store are board games. If the game store sold 256 games in one day, how many of those games were board games?

**18. Banking** Catalina's mother went to the bank to take out $40.00. She asked for 50% of the $40.00 in dollar bills. How much money did she receive in dollar bills?

**19. Standardized Test Practice** What is 30% of 90?

**A** 27 **B** 30 **C** 33 **D** 24

Answers: **1.** 0.15 **2.** 2.8 **3.** 5 **4.** 26 **5.** 28 **6.** 21 **7.** 96 **8.** 48 **9.** 7 **10.** 9 **11.** 99 **12.** 24 **13.** 32 **14.** 54 **15.** 63 **16.** 64 **17.** 192 **18.** $20.00 **19.** A

NAME ______________________ DATE ____________ PERIOD _____

# 10-8 Estimating with Percents (pages 415–417)

When a problem asks for "about how many," the word *about* tells you that an exact answer is not needed. You can estimate the answer.

| | |
|---|---|
| **Common Equivalents for Percents and Fractions** | Memorizing these common equivalents will help you estimate. Often you can think of money to help you remember these. For example: A quarter is $0.25 which is $\frac{1}{4}$ of a dollar. |

| | | | |
|---|---|---|---|
| $20\% = \frac{1}{5}$ | $25\% = \frac{1}{4}$ | $12\frac{1}{2}\% = \frac{1}{8}$ | $16\frac{2}{3}\% = \frac{1}{6}$ |
| $40\% = \frac{2}{5}$ | $50\% = \frac{1}{2}$ | $37\frac{1}{2}\% = \frac{3}{8}$ | $33\frac{1}{3}\% = \frac{1}{3}$ |
| $60\% = \frac{3}{5}$ | $75\% = \frac{3}{4}$ | $62\frac{1}{2}\% = \frac{5}{8}$ | $66\frac{2}{3}\% = \frac{2}{3}$ |
| $80\% = \frac{4}{5}$ | $100\% = 1$ | $87\frac{1}{2}\% = \frac{7}{8}$ | $83\frac{1}{3}\% = \frac{5}{6}$ |

## EXAMPLES

**A** Estimate 61% of 35.

*The table shows that 60% is $\frac{3}{5}$. Multiply to estimate.*

*$\frac{3}{5} \times 35 = 21$. So 61% of 35 is about 21.*

**B** Estimate 9% of 415.

*10% is $\frac{1}{10}$. Multiply to estimate.*

*$\frac{1}{10} \times 415 = 41.5$. So 9% of 415 is about 41.*

### Try These Together

**1.** Estimate 88% of 64.

*HINT: Multiply to find $\frac{7}{8}$ of 64.*

**2.** Estimate 17% of 24.

*HINT: Multiply to find $\frac{1}{6}$ of 24.*

## PRACTICE

***Estimate each percent.***

**3.** 26% of 40

**4.** 18% of 10

**5.** 48% of 30

**6.** 60% of 21

**7.** 73% of 104

**8.** 80% of 51

**9.** 101% of 41

**10.** 34% of 9

**11.** About how much is 48% of 12?

**12.** **School** There are 23 students in Donovan's class. About 25% of his classmates are older than him. Estimate how many of Donovan's classmates are older than him.

**13.** **Standardized Test Practice** Tyler's family gets a busy signal 21% of the time they try to log on to the Internet. If they tried to log on 10 times in one day, about how many times would they get a busy signal?

**A** 2 **B** 3 **C** 4 **D** 5

**Answers:** Sample answers are given. **1.** 56 **2.** 4 **3.** 10 **4.** 2 **5.** 15 **6.** 12 **7.** 75 **8.** 40 **9.** 41 **10.** 3 **11.** 6 **12.** 6 **13.** A

10

# Chapter 10 Review

## Ratio Treasure

Use the treasure map to answer the following questions.

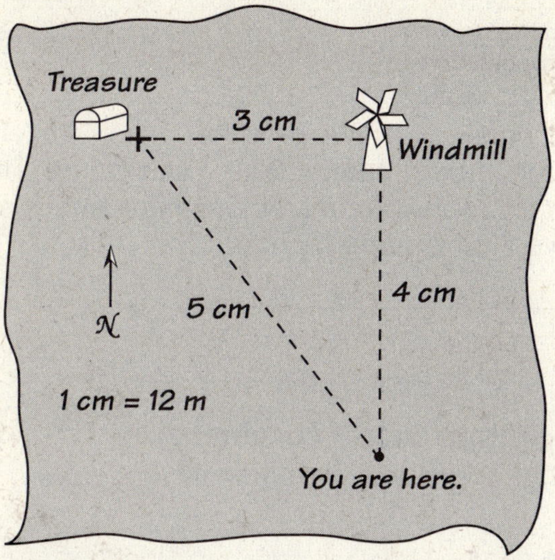


**1.** You're using the map to find a hidden treasure. If you walk directly to the treasure, how far will you walk?

**2.** To make sure you find the treasure, you decide to use a compass to walk north to the windmill first, then east to the treasure. How far are you from the windmill? How far is the windmill from the treasure?

**3.** Suppose instead that you are 60 meters south of a boulder, and the boulder is 80 meters west of a buried treasure. Draw a treasure map with a scale of 1 cm = 20 m. Be sure to label distances on your map according to the scale.

Answers are located on p. 107.

NAME ______________________ DATE ____________ PERIOD _____

# 11-1 Theoretical Probability (pages 428–431)

**Theoretical probability** is the ratio of the number of ways an event can occur to the number of possible outcomes.

| | |
|---|---|
| **Finding Theoretical Probability** | $P(\text{event}) = \dfrac{\text{number of favorable outcomes}}{\text{number of possible outcomes}}$ |
| **Complementary Events** | Complementary events are two events in which either one or the other must take place, but they cannot both happen at the same time. The sum of their probabilities is 1. An example of complementary events is rolling an even or odd number when you roll a number cube. $P(\text{event}_1) + P(\text{event}_2) = 1$ |

## EXAMPLE

A student council representative is to be chosen from a class containing 12 boys and 16 girls. What is the probability that a girl will be chosen?

$\frac{16}{28}$ ← *number of ways to choose a girl* / *number of possible representatives in the class*

*Therefore, P(a girl being chosen to be on the student council)* $= \frac{16}{28}$ *or* $\frac{4}{7}$*, 0.57, or 57%.*

## Try These Together

***There are 5 equally likely outcomes on a spinner, numbered 1, 2, 3, 4, and 5.***

**1.** Find $P$(even number) for the spinner.
*HINT: How many outcomes are even numbers, compared to the total number of outcomes?*

**2.** Find $P$(odd number) for the spinner.
*HINT: How many outcomes are odd numbers, compared to the total number of outcomes?*

## PRACTICE

***A number cube is marked with 1, 2, 3, 4, 5, and 6 on its faces. Suppose you roll the number cube one time. Find the probability of each event. Write each answer as a fraction, a decimal, and a percent.***

**3.** $P(4)$

**4.** $P(4, 5, \text{or } 6)$

**5.** $P(3 \text{ or } 5)$

**6.** $P(1, 2, \text{or } 3)$

**7. Standardized Test Practice** On a science test, 75% of the students got Bs. What is the probability that a particular student did not get a B?

**A** 25% **B** 10% **C** 50% **D** 75%

Answers: **1.** $\frac{2}{5}$, 0.4, 40% **2.** $\frac{3}{5}$, 0.6, 60% **3.** $\frac{1}{6}$, $0.1\overline{6}$, $16.\overline{6}$% **4.** $\frac{1}{2}$, 0.5, 50% **5.** $\frac{1}{3}$, $0.\overline{3}$, $33.\overline{3}$% **6.** $\frac{1}{2}$, 0.5, 50% **7.** A

# 11-2 Outcomes (pages 433–436)

To find outcomes when you are given choices, you can simply list all of the possible outcomes, or use one of the methods below. The set of all possible outcomes is called the **sample space**.

| | |
|---|---|
| **Tree Diagram** | **Size** / **Topping** / **Outcome**<br>small — none → small hot dog without chili<br>small — chili → small hot dog with chili<br>large — none → large hot dog without chili<br>large — chili → large hot dog with chili |
| **Combinations** | **Combinations** are arrangements or listings in which order is not important. To find combinations, make a list. For example, let S stand for a small hot dog, and L stand for a large hot dog, and C stand for chili, and N stand for none. Now, list all of the ways you can pair these letters.<br>SN, NS, SC, CS, LN, NL, LC, CL<br>Since SN and NS are the same, a small hot dog with no chili, then this arrangement is a combination. The four different combinations are SN, SC, LN, and LC. |

## EXAMPLE

At a concession stand, you can order a small, medium, or large cola, with or without ice. Use a tree diagram to find the number of possible outcomes.

| Ice | Size | Outcome |
|---|---|---|
| ice | small | small cola with ice |
| | medium | medium cola with ice |
| | large | large cola with ice |
| no ice | small | small cola without ice |
| | medium | medium cola without ice |
| | large | large cola without ice |

### Try This Together

1. At the school snack bar, you can get apple, grape, or orange juice in a can, bottle, or drink box. Use a tree diagram to find the number of possible outcomes.

## PRACTICE

***Draw a tree diagram to show the sample space for each situation. Then tell how many outcomes are possible.***

2. a choice of black or brown shoes with tan or blue pants
3. a choice of grape, apple, or orange juice with a sandwich or slice of pizza
4. **Standardized Test Practice** The Ramirez family is getting 2 new sofas. In how many ways can they choose 2 sofas from 6 sofas?

**A** 25 **B** 10 **C** 15 **D** 30

**Answers:** **1.** 9 possible outcomes **2.** 4 **3.** 6 For Exercises 1–3, also see students' work to check tree diagrams. **4.** C

NAME ______________________________ DATE ______________ PERIOD _____

# 11-3 Making Predictions (pages 438–441)

If you want to make a prediction about a large group of people, you may wish to use a smaller group, or **sample**, from the larger group. The larger group from which you gathered your sample is known as the **population**. To make sure your information represents the population, the sample must be **random**, or drawn by chance from the population. You can then use the information from the sample to make a prediction about the larger population.

## EXAMPLES

***Kwame found that 20 of the 50 students he surveyed in the lunch line liked enchiladas the best.***

**A** What is the probability that any student will like enchiladas the best?

*20 out of 50, or $\frac{2}{5}$, like enchiladas.*

*The probability is $\frac{2}{5}$, or 40%.*

**B** There are 520 students at Kwame's middle school. Predict how many like enchiladas the best.

*Use a proportion. Let s represent the number of students who like enchiladas. Remember 20 out of 50, or $\frac{2}{5}$ of the students like enchiladas.*

$$\frac{2}{5} = \frac{s}{520}$$

$1{,}040 = 5s$ *Multiply to find the cross products.*

$208 = s$ *Divide each side by 5.*

*About 208 of 520 students probably like enchiladas.*

### Try These Together

***Kwame found that 10 out of the 50 students liked hamburgers the best.***

**1.** What is the probability that any student will like hamburgers the best?
*Hint: Write a ratio.*

**2.** Predict how many of the 520 students will like hamburgers the best.
*Hint: Use a proportion.*

## PRACTICE

**3. Recreation** Carmelina conducted a survey to find out if students preferred in-line skating or skateboarding. 64 out of 80 students preferred in-line skating. There are 200 students at her school. Predict how many of them prefer in-line skating.

**4. Standardized Test Practice** A survey was conducted to find out if people preferred cheddar cheese or mozzarella cheese. 5 out of 20 people preferred cheddar cheese. What is the probability that any given person will prefer cheddar cheese?

**A** $\frac{2}{5}$ **B** $\frac{1}{5}$ **C** $\frac{2}{3}$ **D** $\frac{1}{4}$

**Answers:** **1.** $\frac{1}{5}$ or 20% **2.** about 104 **3.** about 160 **4.** D

NAME ______________________ DATE __________ PERIOD ____

# 11-4 Probability and Area (pages 444–447)

The probability of hitting the bull's-eye in darts is equal to the ratio of the area of the bull's-eye to the total area of the dartboard.

| **Relationship Between Probability and Area** | Suppose you throw a large number of darts at a dartboard. $\frac{\text{number landing in the bull's-eye}}{\text{number landing in the dartboard}} = \frac{\text{area of the bull's-eye}}{\text{total area of the dartboard}}$ |
|---|---|

## EXAMPLES

***A dartboard has three regions, A, B, and C. Region B has an area of 8 in² and regions A and C each have an area of 10 in².***

**A** What is the probability of a randomly thrown dart hitting region B?

$P(\text{region } B) = \frac{\text{area of region B}}{\text{total area of the dartboard}}$

$= \frac{8}{28} \text{ or } \frac{2}{7}$

**B** If you threw a dart 105 times, how many times would you expect it to hit region B?

*Let b = times the dart lands in region B.*

$\frac{b}{105} = \frac{2}{7}$

$7b = 210$ *Multiply to find the cross products.*

$b = 30$ *Divide each side by 7.*

*Out of 105 times, you would expect to hit region B about 30 times.*

## Try These Together

***On a dartboard, region A has an area of 5 in² and region B has an area of 95 in².***

**1.** What is the probability of a randomly-thrown dart hitting region A?

**2.** If you threw a dart 400 times, how many times would you expect it to hit region A?

## PRACTICE

***Each figure represents a dartboard. It is equally likely that a dart will land anywhere on the dartboard. Find the probability of a randomly-thrown dart landing in the shaded region. How many of 100 darts thrown would hit each shaded region?***

**3.** 

**4.** 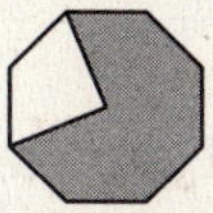

**5.** 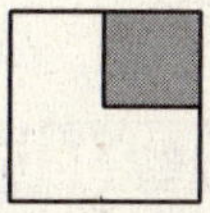

**6.** **Standardized Test Practice** About $\frac{2}{3}$ of the ground under an apple tree is covered with grass, and the rest with dirt. It is equally likely that an apple will fall anywhere on the ground. What is the probability that it will fall on dirt?

**A** $\frac{4}{7}$ **B** $\frac{1}{3}$ **C** $\frac{3}{5}$ **D** $\frac{3}{6}$

**Answers: 1.** $\frac{1}{20}$ **2.** about 20 **3.** $\frac{1}{2}$; about 50 **4.** $\frac{3}{4}$; about 75 **5.** $\frac{1}{4}$; about 25 **6.** B

NAME ______________________ DATE ____________ PERIOD _____

# 11-5 Probability of Independent Events (pages 450–453)

If the outcome of one event does not affect the outcome of another event, the events are called **independent events**.

| Probability of Two Independent Events | The probability of two independent events is the product of the probability of the first event and the probability of the second event. *P*(first event and second event) = *P*(first event) · *P*(second event) |
|---|---|

## EXAMPLES

**A** What is the probability of rolling two 3s in a board game?

$P(3) = \frac{1}{6}$

$P(\text{double 3s}) = P(3) \cdot P(3)$

$= \frac{1}{6} \cdot \frac{1}{6}$

$= \frac{1}{36}$ *Multiply.*

*The probability of rolling double 3s is* $\frac{1}{36}$.

**B** What is the probability of tossing a coin two times and getting heads both times?

$P(\text{tossing heads once}) = \frac{1}{2}$

$P(\text{tossing heads twice}) = \frac{1}{2} \cdot \frac{1}{2}$

$= \frac{1}{4}$ *Multiply.*

*The probability of tossing a coin two times and getting heads both times is* $\frac{1}{4}$.

## Try These Together

**1.** You have two bags. Each contains a yellow, blue, green, and red marble. What is the probability of choosing a blue marble from each bag?
*Hint: Find the probability of each event. Then multiply.*

**2.** With the same bags as Exercise 1, what is the probability of choosing either a yellow or green out of each bag?
*Hint: Find the probability of each event. Then multiply.*

## PRACTICE

***One of 4 different colored balls is chosen from a bag and a number cube is rolled. Find the probability of each event.***

**3.** *P*(red and 2)

**4.** *P*(green and 1 or 2)

**5.** **Standardized Test Practice** Danika and Chantal each have identical boxes of crayons that contain eight different crayons each. What is the probability that they will both pick red when they each pull a crayon out of their boxes?

**A** $\frac{1}{64}$ **B** $\frac{1}{16}$ **C** $\frac{1}{6}$ **D** $\frac{1}{24}$

Answers: 1. $\frac{1}{16}$ 2. $\frac{1}{4}$ 3. $\frac{1}{24}$ 4. $\frac{1}{12}$ 5. A

NAME ______________________________ DATE ____________ PERIOD _____

# 11 Chapter 11 Review

## Board Game Probability

Use what you know about probability to help yourself in this board game against a family member. To move your game pieces, you each roll a standard number cube.

**1.** On the board game above, your game piece is represented by the square and your family member's game piece is represented by the triangle. To win the game, you need to land exactly on the finish square. If you and your family member each roll once, which one of you is more likely to land exactly on the finish square? Explain.

**2.** You hold a card that says if you roll a 6 twice in a row you automatically win. What is the probability that you will roll a 6 twice in a row?

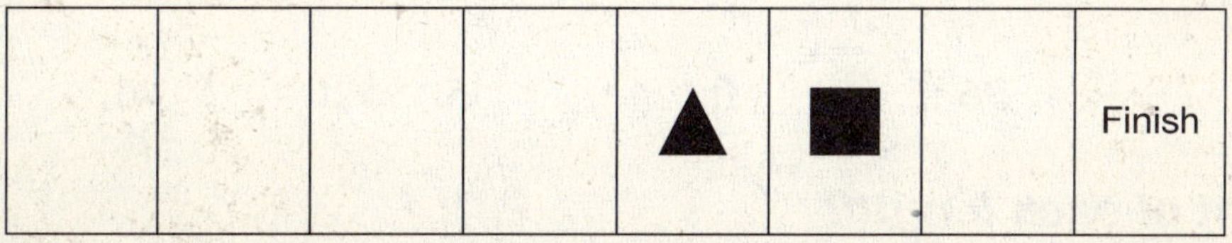


**3.** After one roll each, you and your family member are in the spaces above. What is the probability that you both land in the finish square on the next roll?

Answers are located on p. 107.

NAME ______________________ DATE __________ PERIOD ____

# 12-1 Length in the Customary System (pages 465–468)

Sometimes you need to measure objects using fractions of customary units. The most commonly used customary units of length are the **inch**, **foot**, **yard**, and **mile**.

| | |
|---|---|
| **Customary Units of Length** | 1 foot (ft) = 12 inches (in.)<br>1 yard (yd) = 3 feet or 36 inches<br>1 mile (mi) = 1,760 yards or 5,280 feet |
| **Using a Ruler** | 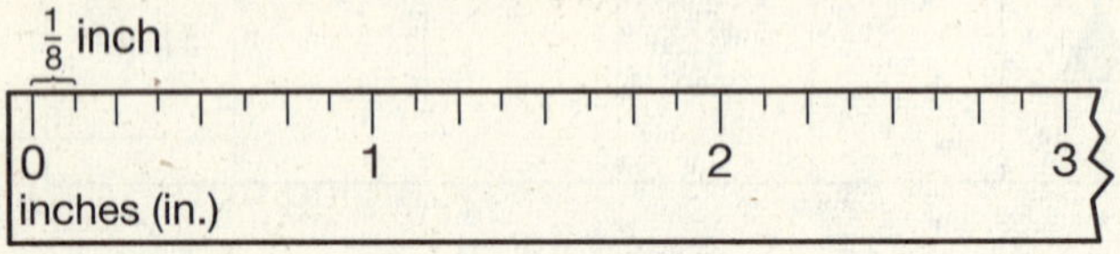<br>Most rulers are separated into eighths. |

## EXAMPLES

**A** 36 in. = __?__ ft

*Since 1 ft = 12 in., it follows that 36 in., or 3 × 12 in., equals 3 ft.*

**B** Draw a line segment measuring $1\frac{3}{8}$ inches.

*Find $1\frac{3}{8}$ on the ruler.*

*Draw a line segment from 0 to $1\frac{3}{8}$.*

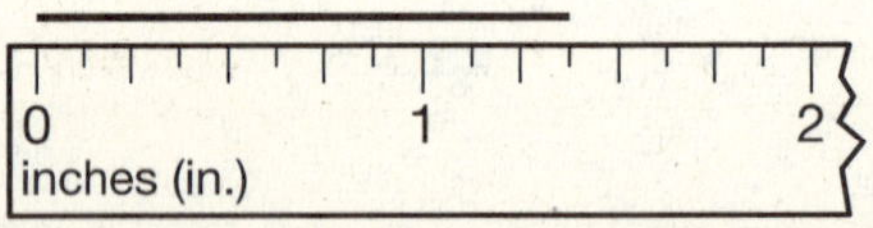


### Try These Together

**1.** 2 mi = __?__ yd

*HINT: Start with 1 mi = 1, 760 yd. Multiply.*

**2.** Draw a line segment measuring $2\frac{1}{4}$ in.

*HINT: How many eighths are in $\frac{1}{4}$?*

## PRACTICE

***Complete.***

**3.** 6 ft = __?__ yd

**4.** 96 in. = __?__ ft

**5.** 36 ft = __?__ yd

***Draw a line segment of each length.***

**6.** $\frac{3}{4}$ inch

**7.** $1\frac{1}{8}$ inches

**8.** $2\frac{3}{8}$ inches

**9. Architecture** A room is 12 feet wide. How many inches wide is the room?

**10. Standardized Test Practice** Complete 9 yd = __?__ in.

**A** 324 **B** 27 **C** 108 **D** 3

**Answers: 1.** 3,520 **2.** See Answer Key. **3.** 2 **4.** 8 **5.** 12 **6–8.** See Answer Key. **9.** 144 in. **10.** A

NAME ______________________ DATE ____________ PERIOD _____

# 12-2 Capacity and Weight in the Customary System (pages 470–473)

The table below lists the most commonly used customary units, and the information you need in order to change them from one unit to another.

| | |
|---|---|
| **Customary Units of Capacity** | 1 cup (c) = 8 fluid ounces (fl oz)<br>1 pint (pt) = 2 cups<br>1 quart (qt) = 2 pints<br>1 gallon (gal) = 4 quarts |
| **Customary Units of Weight** | 1 pound (lb) = 16 ounces (oz)<br>1 ton (T) = 2,000 pounds |
| **Changing Customary Units of Capacity and Weight** | • Determine whether you are changing from smaller to larger units or from larger to smaller units.<br>• To change from smaller to larger units, divide. To change from larger to smaller units, multiply. |

## EXAMPLES

**A** 3 qt = __?__ pt

*Think: Each quart equals 2 pints.*

*3 × 2 = 6* — *Multiply to change from a larger unit (qt) to a smaller unit (pt).*

*3 qt = 6 pt*

**B** 8 c = __?__ qt

*Think: Each quart equals 2 pints and each pint equals 2 cups. You need to divide twice.*

*8 ÷ 2 = 4* — *Divide to change from cups to pints.*

*4 ÷ 2 = 2* — *Divide to change from pints to quarts.*

*8 c = 2 qt*

## Try These Together

**1.** 6 T = __?__ lb

*HINT: You are changing from larger to smaller units.*

**2.** 48 fl oz = __?__ pt

*HINT: You are changing from smaller to larger units.*

## PRACTICE

***Complete.***

**3.** 4 qt = __?__ pt
**4.** 18 fl oz = __?__ c
**5.** 4 gal = __?__ qt
**6.** 8 qt = __?__ c
**7.** 36 oz = __?__ lb
**8.** 5 lb = __?__ oz
**9.** 10 T = __?__ lb
**10.** 17 pt = __?__ gal
**11.** 16 qt = __?__ pt

**12. Standardized Test Practice** An ice cream sundae has 1 cup of ice cream. How many gallons of ice cream would you need to make 64 ice cream sundaes?

**A** 4 gal **B** 2 gal **C** 6 gal **D** 8 gal

Answers: **1.** 12,000 **2.** 3 **3.** 8 **4.** $2\frac{1}{4}$ **5.** 16 **6.** 32 **7.** $2\frac{1}{4}$ **8.** 80 **9.** 20,000 **10.** $2\frac{1}{8}$ **11.** 32 **12.** A

NAME ______________________ DATE ____________ PERIOD ____

# 12-3 Length in the Metric System (pages 476–479)

The basic unit of length in the **metric system** is the **meter**. A centimeter is one-hundredth of a meter. A millimeter is one-thousandth of a meter. A kilometer is a thousand meters.

| | |
|---|---|
| **Choosing a Unit of Length** | A millimeter is about the width of the lead in a pencil.<br>A centimeter is about the width of a little fingernail.<br>A meter is about the length of the handle of a broom.<br>A kilometer is about the length of TEN football fields. |

## EXAMPLES

**A** How many meters are in 5 kilometers?
*One kilometer is 1,000 meters. Two kilometers is 2 × 1,000 or 2,000 meters. There are 5,000 meters in 5 kilometers.*

**B** Use a centimeter ruler to measure the width of a piece of notebook paper.
*The width is about 21.5 centimeters.*

### Try These Together

**1.** What unit of length in the metric system would you use to measure the distance across your city or town?
*HINT: What unit is large enough to use for long distances?*

**2.** What metric unit of length would you use to measure the thickness of a piece of cardboard?
*HINT: Choose a unit that is very small.*

## PRACTICE

***Write the metric unit of length that you would use to measure each of the following.***

**3.** height of a refrigerator

**4.** length of a banana

**5.** thickness of a quarter

**6.** distance from New York to Los Angeles

**7.** length of a car

**8.** height of a two-story house

**9.** How many centimeters are in 2 meters?

**10.** How many meters are in 8 kilometers?

**11.** **School** For a science experiment, students need a piece of string about as long as their science textbook. What metric unit should they use to measure the string?

**12.** **Standardized Test Practice** How long is the peanut in centimeters?

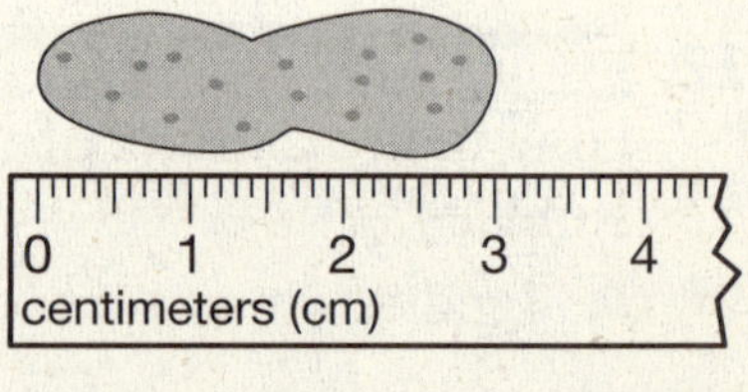


**A** 2 centimeters **B** 3 centimeters
**C** 4 centimeters **D** 5 centimeters

**Answers:** 1. kilometer 2. millimeter 3. meter 4. centimeter 5. millimeter 6. kilometer 7. meter 8. meter 9. 200 10. 8,000 11. centimeter 12. B

NAME ______________________ DATE ____________ PERIOD ______

# 12-4 Mass and Capacity in the Metric System (pages 484–487)

In the metric system, all units are defined in terms of a basic unit. The basic unit of mass is the **kilogram** (kg). The basic unit of capacity is the **liter** (L).

| | | | |
|---|---|---|---|
| **Metric Units of Mass** | **gram** (g)<br>**kilogram** (kg)<br>**milligram** (mg) | 1,000 g = 1 kg<br><br>1 mg = 0.001 g | A small paperclip has a mass of about 1 gram.<br>A textbook has a mass of about 1 kilogram.<br>A grain of salt has a mass of about 1 milligram. |
| **Metric Units of Capacity** | **liter** (L)<br>**milliliter** (mL) | <br>1 mL = 0.001 L | A small pitcher has a capacity of about 1 liter.<br>An eyedropper holds about 1 milliliter of liquid. |

## EXAMPLES

***What unit would you use to measure each of the following?***

**A** the mass of a compact car

*Even a compact car has quite a bit of mass. The kilogram is the appropriate unit to measure the mass of a compact car. The average compact car has a mass of about 1,200 kilograms.*

**B** the capacity of a soda can

*Since a liter is about the same capacity as a quart, you know that a soda can has less than one liter of capacity. The milliliter is the appropriate unit to measure the capacity of a soda can, which holds about 355 mL.*

### Try These Together

***What unit would you use to measure each of the following? Estimate the mass or capacity.***

**1.** a coffee cup

**2.** a candy bar

## PRACTICE

***Write the metric unit of mass or capacity that you would use to measure each of the following. Then estimate the mass or capacity.***

**3.** a wading pool

**4.** a hammer

**5.** the wings of a housefly

**6.** the ink in a fountain pen

**7.** a nickel

**8.** a bird bath

***Name an item that you think has the given measure.***

**9.** about 20 g

**10.** about 500 mL

**11.** about 2 L

**12.** about 5 kg

**13. Food** A bottle of grape juice has a capacity of 1890 mL. If the bottle has eight servings, how many mL is one serving?

**14. Standardized Test Practice** What is the mass of a large can of tomatoes?

**A** 1 mL **B** 1 L **C** 1 g **D** 1 kg

**Answers: 1.** milliliter, about 200 mL **2.** gram, about 50 g **3.** liter, about 1,000 L **4.** kilogram; about 1 kg **5.** milligram; about 2 mg **6.** milliliter; about 1 mL **7.** gram; about 5 g **8.** liter; about 8 L **9–12.** Answers will vary. **13.** 236.25 mL **14.** D

# 12-5 Changing Metric Units (pages 490–493)

To change from one metric unit to another, you either multiply or divide by powers of 10. The chart below shows the relationship between the metric units and the powers of 10.

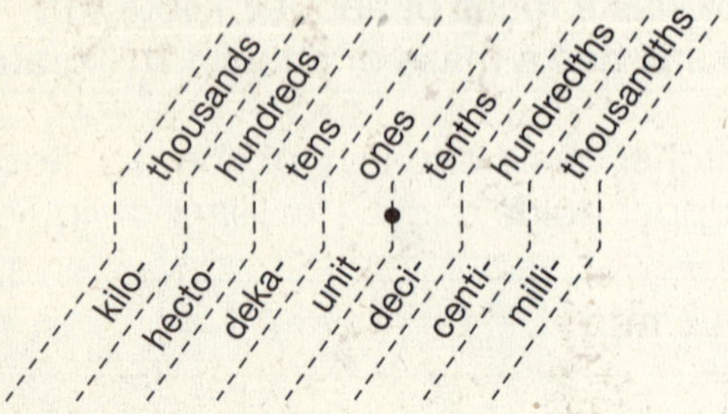


To change from a larger unit to a smaller unit, you need to multiply. To change from a smaller unit to a larger unit, you need to divide.

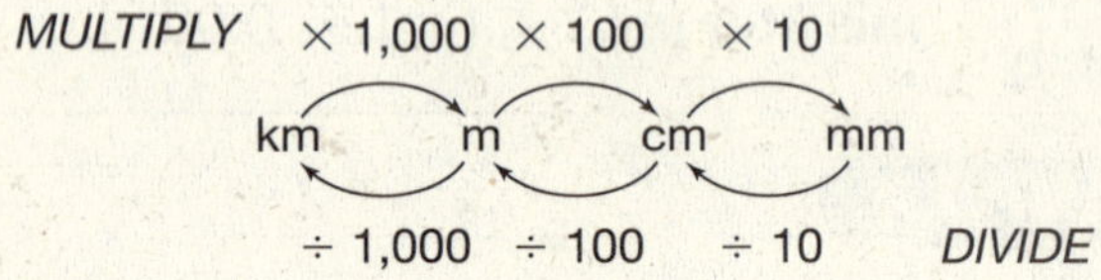


## EXAMPLES

**A** 1.5 L = __?__ mL

*To change from liters to milliliters, multiply by 1,000 since 1 mL = 0.001 L.*

*1.5 × 1,000 = 1,500*

*1.5 L = 1,500 mL*

**B** 12 cm = __?__ m

*To change from centimeters to meters, divide by 100 since 1 m = 100 cm.*

*12 ÷ 100 = 0.12*

*12 cm = 0.12 m*

## Try These Together

***Complete.***

**1.** 3 kg = __?__ g

*HINT: Kilograms are larger units than grams; multiply.*

**2.** 9 mm = __?__ cm

*HINT: Millimeters are smaller units than centimeters; divide.*

## PRACTICE

***Complete.***

**3.** 4,860 mm = __?__ km
**4.** __?__ L = 397 mL
**5.** 669 mm = __?__ cm
**6.** __?__ mg = 0.0079 g
**7.** 8,170 mm = __?__ m
**8.** __?__ mL = 7.6 L
**9.** 0.0034 kg = __?__ mg
**10.** __?__ mg = 0.4 g
**11.** 460 mL = __?__ L
**12.** __?__ g = 557 mg
**13.** 748 cm = __?__ m
**14.** __?__ mL = 0.06 L
**15.** 1.68 km = __?__ cm
**16.** __?__ g = 8.05 kg
**17.** 336 m = __?__ km

**18. Food** A baby drinks 85 milliliters of juice a day. How many liters of juice does the baby drink in a week?

**19. Standardized Test Practice** How many centimeters are in 0.082 kilometers?

**A** 8.2 **B** 82 **C** 8,200 **D** 82,000

**Answers: 1.** 3,000 **2.** 0.9 **3.** 0.00486 **4.** 0.397 **5.** 66.9 **6.** 7.9 **7.** 8.17 **8.** 7,600 **9.** 3,400 **10.** 400 **11.** 0.46 **12.** 0.557 **13.** 7.48 **14.** 60 **15.** 168,000 **16.** 8,050 **17.** 0.336 **18.** 0.595 L **19.** C

NAME ______________________ DATE ____________ PERIOD _____

# 12-6 Measures of Time (pages 494–497)

You add and subtract measures of time in the same way you add and subtract mixed numbers.

| | |
|---|---|
| **Adding and Subtracting Measures of Time** | • Add or subtract the seconds.<br>• Add or subtract the minutes.<br>• Finally, add or subtract the hours.<br>Rename, if necessary, in each step. |

## EXAMPLES

**A** Find 3 h 15 min + 2 h 20 min.

| *First add the minutes.* | | *Then add the hours.* |
|---|---|---|
| *3 h 15 min* | | *3 h 15 min* |
| *+ 2 h 20 min* | → | *+ 2 h 20 min* |
| *35 min* | | *5 h 35 min* |

**B** Find 8 h 12 min − 6 h 48 min.

| *First rename.* | | *Subtract the minutes.* | | *Subtract the seconds.* |
|---|---|---|---|---|
| *7 h 72 min* | | *7 h 72 min* | | *7 h 72 min* |
| *− 6 h 48 min* | → | *− 6 h 48 min* | → | *− 6 h 48 min* |
| | | *24 min* | | *1 h 24 min* |

### Try These Together

***Add or subtract.***

**1.** 4 min 32 s + 8 min 41 s
*HINT: Add the seconds, and then add the minutes.*

**2.** 11 min 4 s − 5 min 12 s
*HINT: Rename, subtract the seconds, and then subtract the minutes.*

## PRACTICE

***Complete.***

**3.** 3 h 14 min = 2 h __?__ min

**4.** 17 h 18 min = 16 hr __?__ min

**5.** 12 h 6 min = 11 hr __?__ min

**6.** 2 h 9 min 62 s = 2 h __?__ min 2 s

***Add or subtract.***

**7.** 8 h 46 min − 1 h 52 min

**8.** 4 h 36 min − 3 h 5 min

**9.** 6 h 24 min + 4 h 18 min

**10.** 5 h 43 min 21 s − 2 h 18 min 14 s

**11.** 1 h 12 min 36 s + 8 h 54 min 4 s

**12.** 7 h 42 min 16 s + 1 h 58 min 12 s

**13.** **Standardized Test Practice** Margarita is flying from Chicago to Denver. Her 2 h 35 min flight leaves Chicago at 5:55 P.M. What time does the flight arrive in Denver? Hint: The local Chicago time is one hour ahead of the local time in Denver.

**A** 6:30 P.M. **B** 5:30 P.M. **C** 7:30 P.M. **D** 8:30 P.M.

**Answers:** **1.** 13 min 13 s **2.** 5 min 52 s **3.** 74 **4.** 78 **5.** 66 **6.** 10 **7.** 6 h 54 min **8.** 1 h 31 min **9.** 10 h 42 min **10.** 3 h 25 min 7 s **11.** 10 h 6 min 40 s **12.** 9 h 40 min 28 s **13.** C

NAME ______________________________ DATE ____________ PERIOD _____

# Chapter 12 Review

## Classroom Math

You will need a tape measure for this activity. Round all answers to the nearest hundredth.

**EXAMPLE**

**Find the length of your classroom in yards. How many feet are in 1 yard? Find the length of your classroom in feet.**

**Yards:** 12 yards
**Conversion:** 3 feet in 1 yard, $3 \cdot 12 = 36$
**Feet:** 36 feet

1. Find the height of the door to your classroom in inches. How many inches are in 1 foot? Find the height of the door in feet.

   Inches: _____
   _____ inches in 1 foot
   Feet: _____

2. Find the length of your textbook in centimeters. How many centimeters are in 1 meter? Find the length of your textbook in meters.

   Centimeters: _____
   _____ centimeters in 1 meter
   Meters: _____

3. Estimate the number of miles you live from school. How many feet are in 1 mile? Find the number of feet you live from school.

   Miles: _____
   _____ feet in 1 mile
   Feet: _____

4. Find an object in the classroom that is approximately $1\frac{1}{2}$ inches long. Name this object. How long is this object in feet?

   Object: _____
   Feet: _____

5. Find an object in the classroom that is approximately 3.5 centimeters long. How many millimeters are in 1 centimeter? How long is this object in millimeters?

   Object: _____
   _____ millimeters in 1 centimeter
   Millimeters: _____

6. Find an object in the classroom that is approximately 4 grams. How many ounces are in 1 gram? How much does this object weigh in ounces?

   Object: _____
   _____ ounces in 1 gram
   Ounces: _____

Answers are located on p. 107.

NAME ______________________ DATE ____________ PERIOD _____

# 13-1 Angles (pages 506–509)

The lines that form the **edges** of a box meet at a point called the **vertex**. Two lines that meet at a vertex form an **angle**. Angles are measured in **degrees**, or parts of a circle. A circle contains 360 degrees. You can measure the degrees in an angle with a **protractor**.

| | |
|---|---|
| **Classifying Angles** | • Acute angles measure between 0° and 90°.<br>• Obtuse angles measure between 90° and 180°.<br>• Right angles measure 90°.<br>• Complementary angles are two angles whose measures add to 90°.<br>• Supplementary angles are two angles whose measures add to 180°. |

## EXAMPLES

**A** An angle measures 179°. Is it acute, right, or obtuse?

*This angle measures between 90° and 180°, so it is obtuse.*

**B** Angles $F$ and $G$ are complementary angles. Find $m\angle G$ if $m\angle F$ is 31°.

$m\angle G + m\angle F = 90°$

$m\angle G + 31° = 90°$

$m\angle G = 59°$ $\quad 59 + 31 = 90$

## Try These Together

**1.** An angle measures 29°. Is it acute, right, or obtuse?
*HINT: Is 29° less than 90°?*

**2.** Angles $K$ and $L$ are supplementary angles. Find $m\angle K$ if $m\angle L$ is 42°.
*HINT: What is the sum of $m\angle K$ and $m\angle L$?*

## PRACTICE

***Use a protractor to find the measure of each angle.***

**3.**

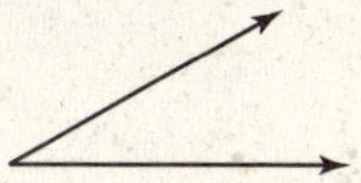

**4.**

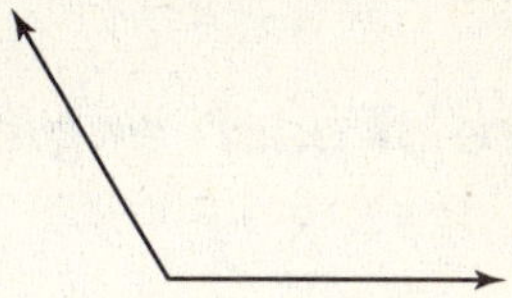

***Classify each angle measure as acute, right, or obtuse.***

**5.** 45° **6.** 100° **7.** 90° **8.** 20°

**9. Architecture** An architect is designing a building. A corner in a hallway has an angle that measures 135°. Is the angle *acute*, *right*, or *obtuse*?

**10. Standardized Test Practice** Angle $P$ and angle $Q$ are complementary. Find $m\angle P$ if $m\angle Q$ is 45°.

**A** 55° **B** 45° **C** 135° **D** 145°

Answers: 1. acute 2. 138° 3. 30° 4. 120° 5. acute 6. obtuse 7. right 8. acute 9. obtuse 10. B

# 13-2 Using Angle Measures (pages 510–512)

You can use a protractor and a **straightedge**—a ruler or any object with a straight side—to draw an angle with a measure of a certain number of degrees. You can also estimate the measure of an angle.

| | |
|---|---|
| **Estimating the Measure of an Angle** | Estimate the measure of an angle by comparing it to a right angle (90°), half of a right angle (45°), one third of a right angle (30°), or two thirds of a right angle (60°). You can also compare an angle to a straight angle (180°). |

**EXAMPLES**

**A** Use two pencils to show an angle of about 35°.

*Think: How does 35° compare to 90°?*

*Hold the pencils to show an angle a little more than one third of a right angle and a little less than half of a right angle.*

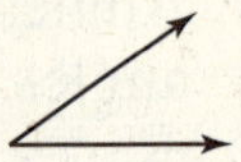

**B** Is this angle greater than, less than, or about equal to 125°?

*The angle shown is just a little less than 180° so it is greater than 125°.*

## Try These Together

1. Use a straightedge to draw an angle that you estimate to be about 22°.
   *HINT: What is half of 45°?*
2. Use a straightedge to draw an angle that you estimate to be about 135°.
   *HINT: Notice that 135° is 90° plus 45°.*

**PRACTICE**

***Use a protractor and a straightedge to draw angles having the following measurements.***

3. 80°  4. 145°  5. 45°  6. 110°

***Estimate the measure of each angle.***

7.
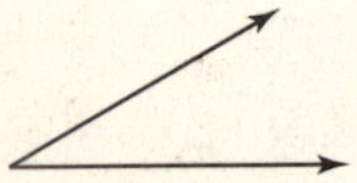

8.
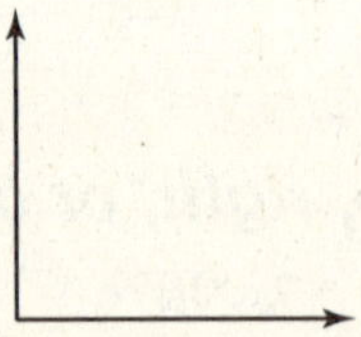

9.
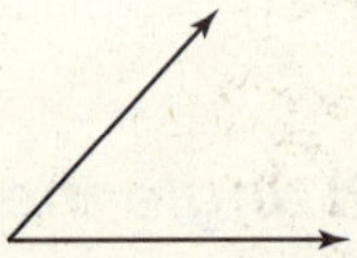

10. **Standardized Test Practice** The circle graph shows what people prefer to eat for breakfast. Which of the following shows the order of breakfasts from most-preferred to least-preferred?

**A** eggs, toast, cereal  **B** cereal, eggs, toast
**C** toast, eggs, cereal  **D** eggs, cereal, toast

**Breakfast Preferences**

**Answers: 1–2.** Use a protractor to see how close your estimates are. **3–6.** See Answer Key. **7–9.** Sample answers are given. **7.** about 30° **8.** about 90° **9.** about 45° **10.** D

NAME ______________________ DATE ____________ PERIOD ____

# 13-3 Bisectors (pages 515–517)

When you **bisect** a geometric figure, you divide it into two congruent parts. A line segment is the **perpendicular bisector** of another line segment when it bisects the segment at a right angle. You can use a straightedge and a compass to bisect a line segment or an angle.

| | |
|---|---|
| **Constructing Bisectors** | • From each end of a line segment, use the same compass setting to draw arcs above and below the line segment. Join the points where the arcs intersect to draw the perpendicular bisector of the segment.<br>• From the vertex of an angle, draw an arc that intersects the sides of the angle. From these two points of intersection, draw equal arcs inside the angle. Join the points where the arcs intersect to the vertex to make a ray that bisects the angle. |

## EXAMPLES

**A** When you draw a ray to bisect an angle of 56°, what is the measure of each angle formed?

*Bisect means to divide into two equal parts, so each angle is one half of 56°, or 28°.*

**B** When you draw the ray that bisects a right angle, are the two angles that result supplementary or complementary?

*Since the two angles total 90°, they are complementary.*

### Try These Together

1. Draw a rectangle that is *not* a square. Draw the two diagonals that connect the opposite corners. Do the diagonals appear to bisect each other?
   *HINT: For each diagonal, compare the lengths of the two parts formed by the point where the diagonals intersect.*

2. Draw a rectangle that is *not* a square. Draw the two diagonals that connect the opposite corners. Is one diagonal the perpendicular bisector of the other?
   *HINT: Measure the angles formed where the diagonals intersect to see if they are 90°.*

## PRACTICE

***Draw each line segment or angle having the given measurement. Then use a straightedge and a compass to bisect the line segment or angle.***

**3.** 90° **4.** 4 cm **5.** 68° **6.** 3 in. **7.** 124°

**8.** **Standardized Test Practice** Angle *FGI* has been bisected by $\overrightarrow{GJ}$. If $m\angle FGI$ is 80°, what is the measure of each angle formed ($\angle FGJ$ and $\angle JGI$)?

**A** 60° **B** 30° **C** 50° **D** 40°

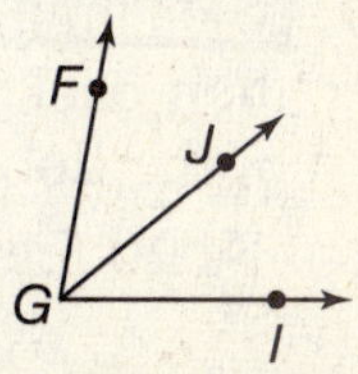


Answers: 1. yes 2. no 3–7. See Answer Key. 8. D

NAME ______________________ DATE ____________ PERIOD ______

# 13-4 Two-Dimensional Figures (pages 522–525)

A polygon with all the sides and angles congruent is called a **regular polygon**. A regular triangle (3 sides) is also called an **equilateral triangle**. In a regular quadrilateral (4 sides), also called a square, the opposite sides are **parallel**. Parallel lines will never meet, no matter how far they are extended.

| | | |
|---|---|---|
| **Identifying Polygons** | A triangle has 3 sides.<br>A pentagon has 5 sides.<br>An octagon has 8 sides. | A quadrilateral has 4 sides.<br>A hexagon has 6 sides.<br>A decagon has 10 sides. |

## EXAMPLES

**A** Is this figure a quadrilateral? Is it a parallelogram?

*Yes, it has 4 sides so it is a quadrilateral. Yes, the opposite sides are parallel, so it is a parallelogram.*

**B** In the figure for Example A, are all the angles congruent? Are the sides? Is this figure a regular polygon?

*Yes, all the angles are right angles so they are congruent. No, the length is greater than the width, so the sides are not congruent and it is not a regular polygon.*

### Try These Together

**1.** How many congruent angles does a regular decagon have?

*HINT: What does "regular" mean? How many sides does a decagon have? Think of the word "decimal" to help you remember the sides of a "decagon."*

**2.** What do you know about a figure if you know that it is a regular hexagon?

*HINT: How many sides does it have? What is true of all the sides and all the angles?*

## PRACTICE

***Identify each polygon. Then tell if it is a regular polygon.***

**3.**

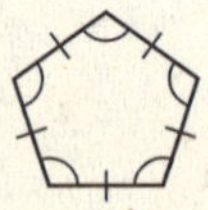

**4.**

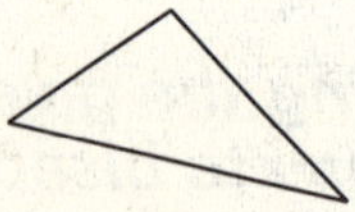

**5.**

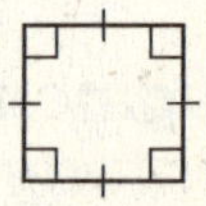

**6.** How many sides does a regular octagon have?

**7.** **Standardized Test Practice** Embry's father is building a storage shed in their backyard. The floor will be the shape of a square. If the perimeter of the floor is 40 feet, how long is each side?

**A** 20 feet **B** 15 feet **C** 10 feet **D** 30 feet

**Answers: 1.** 10 **2.** It has 6 congruent sides and 6 congruent angles. **3.** pentagon; regular **4.** triangle; not regular **5.** square; regular **6.** 8 **7.** C

# 13-5 Lines of Symmetry (pages 528–531)

When a figure has a **line of symmetry** (or more than one), you can fold the figure along this line so that the two halves match. Figures that can be turned or rotated less than 360° about a fixed point and still look exactly the same have **rotational symmetry**.

| **Finding Lines of Symmetry** | To look for lines of symmetry, imagine folding the figure in half vertically, horizontally, and diagonally. When the two halves match exactly, then the fold line is a line of symmetry. |
|---|---|

## EXAMPLES

**A** Draw a line of symmetry for the figure at the right.

*Think about folding the figure along a line to see if the two halves match.*

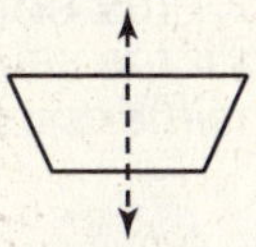

**B** Does the figure in Example A have more than one line of symmetry?

*No. If you draw a diagonal and fold the figure along it, the two halves do not match. The same is true for a line halfway up the figure.*

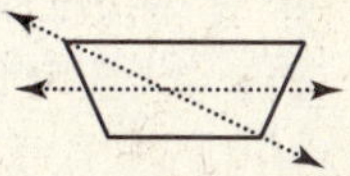

### Try These Together

**1.** How many lines of symmetry does an equilateral triangle have?

*HINT: Sketch the triangle and think about folding it.*

**2.** Do a rectangle (that is not a square) and a square have the same number of lines of symmetry?

*HINT: Look at the diagonals to see if they are lines of symmetry.*

## PRACTICE

***Draw all lines of symmetry in each figure.***

**3.** 

**4.** 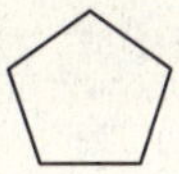

**5.** 

**6.** 

***Tell whether each figure has rotational symmetry. Write yes or no.***

**7.** 

**8.** 

**9.** **Standardized Test Practice** How many lines of symmetry does this shell have?

**A** 1  **B** 2  **C** 3  **D** 4

**Answers: 1.** 3 **2.** no **3–6.** See Answer Key. **7.** no **8.** yes **9.** A

# 13-6 Similar and Congruent Figures (pages 534–536)

You can compare figures that look alike in two different ways.

| | |
|---|---|
| **Comparing Figures for Size and Shape** | Two figures that have the same shape and angles but are different in size are called **similar figures.** Figures that are exactly the same size and shape are called **congruent figures.** |

## EXAMPLES

**A** Is Figure 1 similar or congruent to Figure 2?

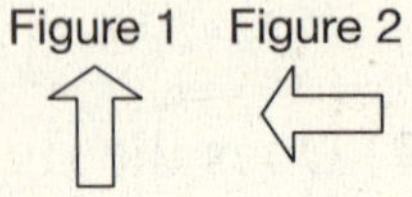


*Although the two figures are turned differently, they are exactly the same size and shape, so they are congruent figures.*

**B** Is Figure 1 similar or congruent to Figure 2?

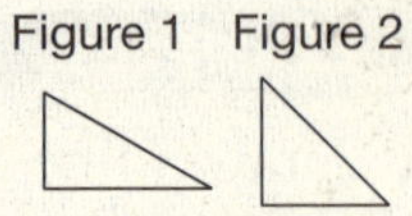


*Although the figures are both right triangles, they are not the same size and they are not the same shape, so they are neither similar nor congruent.*

### Try These Together

**1.** Figure 1 is congruent to Figure 2. Which side of Figure 1 corresponds to side $\overline{MN}$ of Figure 2?

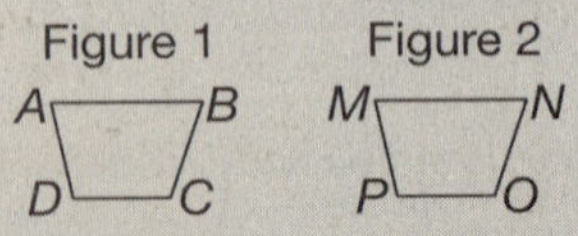


*HINT: Find the side that is in the matching position.*

**2.** Is this pair of polygons *congruent*, *similar*, or *neither*?

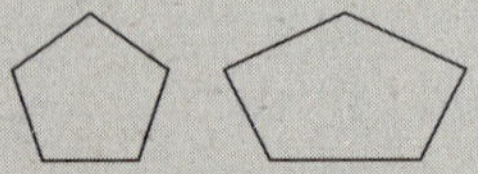

*HINT: Are the figures the same shape? Are they the same size? Are the corresponding angles equal?*

## PRACTICE

***Tell whether each pair of figures is* congruent, similar, *or* neither.**

**3.**

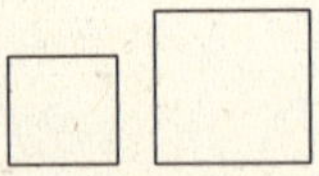

**4.**

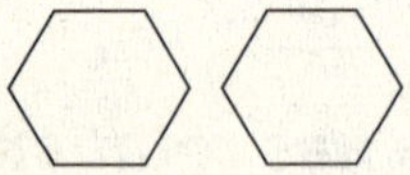

**5.** $\triangle PQR$ is congruent to $\triangle STV$.

**a.** What side corresponds to side $\overline{TV}$?

**b.** What is the measure of side $\overline{PR}$?

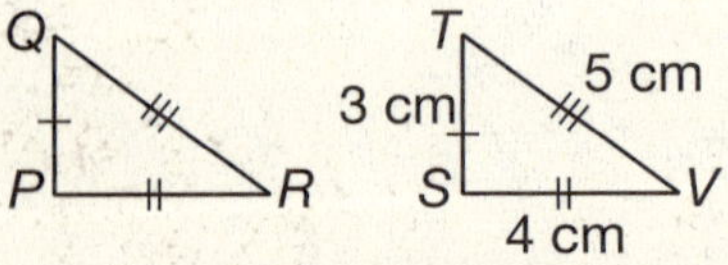


**6. Standardized Test Practice** Which two figures are congruent?

**Answers: 1.** $\overline{AB}$ **2.** neither **3.** similar **4.** congruent **5a.** $\overline{QR}$ **5b.** 4 cm **6.** C

# 13 Chapter 13 Review

## Geometric Combinations

Marta uses the following drawings on pieces of paper to help her remember her locker combination.

**1.** What is Marta's Locker combination? Explain how you know.

**2.** If your locker combination is 48-35-10, make some drawings that could help you remember the combination. (Hint: You can represent 10 with just one polygon.)

Answers are located on p. 108.

NAME ______________________ DATE ____________ PERIOD ______

# 14-1 Area of Parallelograms (pages 546–549)

A parallelogram is a quadrilateral with two pairs of parallel sides. The base is any one of the sides and the height is the shortest distance (the length of a perpendicular segment) from the base to the opposite side.

| **Finding the Area of a Parallelogram** | The area $A$ of a parallelogram equals the product of its base $b$ and height $h$. $A = bh$ |
|---|---|

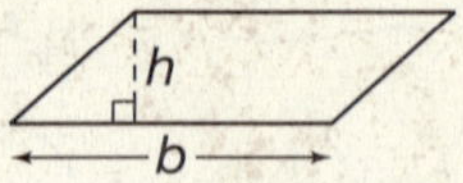


## EXAMPLES

**A** Find the area of the parallelogram.

*Multiply the length of the base of the parallelogram (4 in.) and the height drawn to that base (5 in.).*

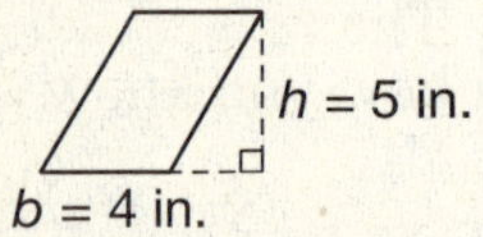


$A = bh$

$A = 4(5) = 20 \text{ in}^2$

**B** The area of a parallelogram is 30 square inches. The base is 10 inches long. What is the height?

$A = bh$

$30 = 10h$    *Substitute the values you know.*

$h = 3$    $30 = 10 \cdot 3$

*The height is 3 inches.*

### Try These Together

1. Find the area (to the nearest tenth) of a parallelogram that is 3.6 centimeters wide and 5.2 centimeters high.
   *HINT: Use the formula and then round.*

2. Find the base of a parallelogram that has a height of 7 centimeters and an area of 56 square centimeters.
   *HINT: Write the formula, substitute values, and solve for b.*

## PRACTICE

***Find the area of each parallelogram.***

3.

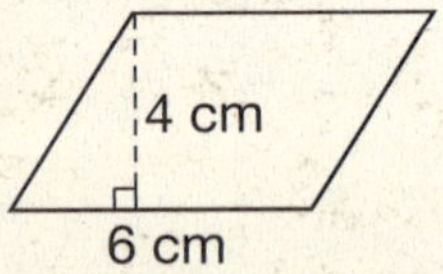


4.

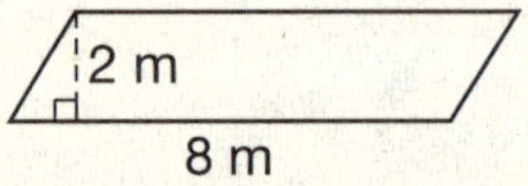


5.

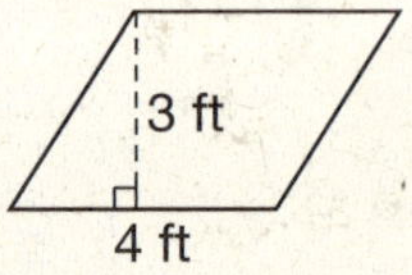


6. What is the area of a parallelogram that is 5 centimeters wide and 8 centimeters high?

7. **Puzzles** Kai has a puzzle that is a parallelogram. It is 30 centimeters long and 22 centimeters high. What is the area of the puzzle?

8. **Standardized Test Practice** If a parallelogram has an area of 42 square centimeters and its height is 6 centimeters, how long is its base?

**A** 6 cm    **B** 7 cm    **C** 5 cm    **D** 8 cm

Answers: **1.** 18.7 in$^2$ **2.** 8 cm **3.** 24 cm$^2$ **4.** 16 m$^2$ **5.** 12 ft$^2$ **6.** 40 cm$^2$ **7.** 660 cm$^2$ **8.** B

NAME ______________________ DATE ____________ PERIOD _____

# 14-2 Area of Triangles (pages 551–554)

You can divide a parallelogram into two congruent triangles by drawing a diagonal. Since the formula for the area of a parallelogram is $A = bh$, then the formula for the area of a triangle is $A = \frac{1}{2}bh$.

| | | |
|---|---|---|
| **Finding the Area of a Triangle** | The area $A$ of a triangle equals half of the product of the length of the base $b$ and the height $h$. $A = \frac{1}{2}bh$ | 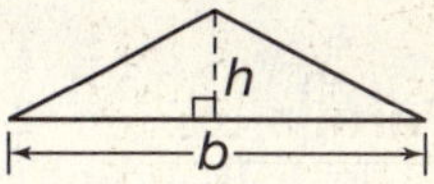 |



## EXAMPLES

**A** What is the area of a triangle with a height of 25 cm and a base of 36 cm?

$A = \frac{1}{2}bh$ *Write the formula.*

$A = \frac{1}{2}(36)(25)$ *Substitute the values you know.*

$A = 450 \text{ cm}^2$ *Multiply to find the area.*

**B** The area of a triangle is 54 in$^2$ and the height is 12 in. Find the base.

$A = \frac{1}{2}bh$ *Write the formula.*

$54 = \frac{1}{2}(b)(12)$ *Substitute the values you know.*

$54 = 6b$ *Multiply.*

$9 \text{ in.} = b$ *$54 = 6 \cdot 9$*

### Try These Together

**1.** Find the area of a triangle that has a base of 1 yd and height of $\frac{1}{3}$ yd.
*HINT: Use the formula and multiply.*

**2.** A triangle has a base of 8 cm and an area of 64 cm$^2$. Find the height.
*HINT: Substitute in the formula and solve for h.*

## PRACTICE

***Find the area of each triangle.***

**3.**
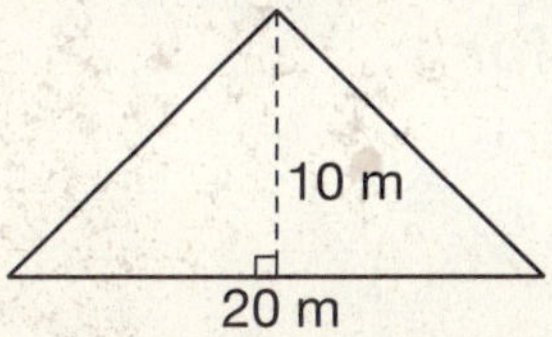


**4.**
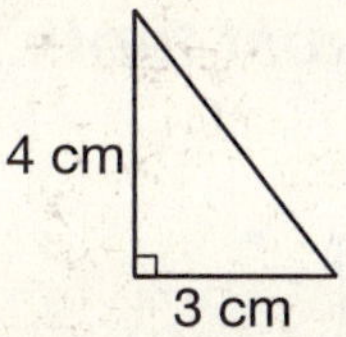


**5.**
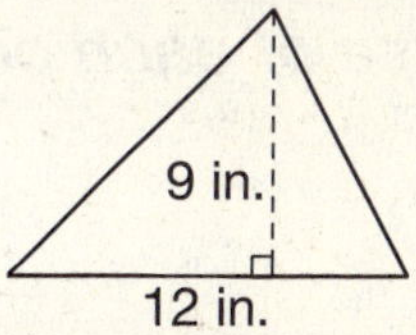


**6. Flags** The flag of the country of Guyana has a red triangle on it. If the base of the triangle is 30 inches and the height is 26 inches, what is the area of the triangle?

**7.** **Standardized Test Practice** How long is the base of a triangle that has an area of 63 square centimeters and a height of 7 centimeters?

**A** 7 cm **B** 9 cm **C** 16 cm **D** 18 cm

**Answers:** **1.** $\frac{1}{6}$ yd$^2$ **2.** 16 cm **3.** 100 m$^2$ **4.** 6 cm$^2$ **5.** 54 in$^2$ **6.** 390 in$^2$ **7.** D

NAME ______________________ DATE ____________ PERIOD _____

# 14-3 Area of Circles (pages 556–559)

If you cut a circle into a number of equal-sized pie-shaped pieces and arrange them carefully, you can form a rough parallelogram. The height of the parallelogram is about equal to the radius of the circle. The base is about equal to $\frac{1}{2}$ of the circumference of the circle. This would mean that the area is about $\frac{1}{2}Cr$. Substitute the circumference formula for $C$ and you get the following equation.

| Finding the Area of a Circle | The area $A$ of a circle equals the product of $\pi$ and the square of the radius $r$. $A = \pi r^2$ |
|---|---|

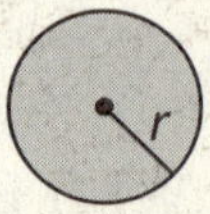


## EXAMPLES

**A** Find the area of a circle with a radius of 7 cm. Use 3.14 for $\pi$.

$A = \pi r^2$ *Write the formula.*

$A \approx 3.14(7)^2$ *Substitute the values you know.*

$A \approx 154\ cm^2$ *Use a calculator and round.*

**B** Find the area of a circle that has a diameter of 5 inches. Use 3.14 for $\pi$.

$A = \pi r^2$ *Write the formula.*

$A \approx 3.14(2.5)^2$ $r = \frac{1}{2}d$ *or 2.5 in.*

$A \approx 19.6\ in^2$ *Use a calculator and round.*

### Try These Together

1. A circle has a radius of 2 in. What is its area? Use 3.14 for $\pi$.
   *HINT: Write the formula and substitute.*

2. The diameter of a circle is 4.2 yd. Find its area. Use 3.14 for $\pi$.
   *HINT: First find the radius.*

## PRACTICE

***Find the area of each circle to the nearest tenth. Use 3.14 for $\pi$.***

3.
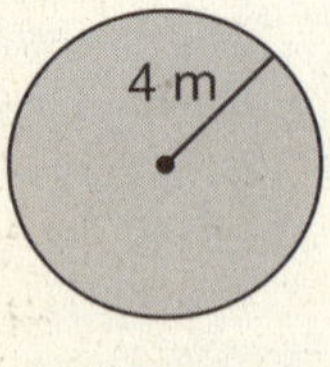


4.
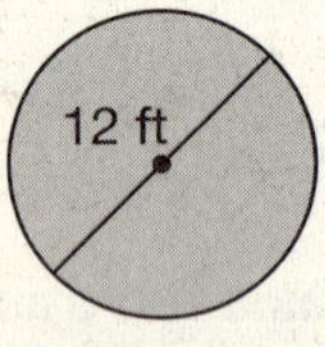


5.

6. diameter, 18 centimeters
7. radius, 5 meters
8. radius, 10 inches

9. **Standardized Test Practice** What is the area of a circle that has a diameter of 30 centimeters?

**A** 353.3 $cm^2$ **B** 2,826 $cm^2$ **C** 176.6 $cm^2$ **D** 706.5 $cm^2$

**Answers: 1.** about 12.6 $in^2$ **2.** about 13.8 $yd^2$ **3.** 50.2 $m^2$ **4.** 113.0 $ft^2$ **5.** 28.3 $in^2$ **6.** 254.3 $cm^2$ **7.** 78.5 $m^2$ **8.** 314.0 $in^2$ **9.** D

NAME ______________________ DATE __________ PERIOD ____

# 14-4 Three-Dimensional Figures (pages 564–566)

A **three-dimensional** figure encloses a part of space. The flat surfaces are called **faces**. The segments formed by the intersecting faces are the **edges**. The edges intersect at the **vertices**.

| | |
|---|---|
| **Identifying Three-Dimensional Figures** | • **prism:** two parallel and congruent faces, called **bases**<br>• **pyramid:** triangular faces; one base<br>Prisms and pyramids are named by the polygon(s) at their base(s).<br>• **cone:** curved surface; one circular base<br>• **cylinder:** curved surface; two circular bases<br>• **sphere:** all the points are the same distance from the center |

## EXAMPLES

**A** Identify this figure.

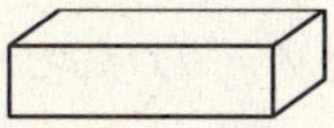

*The faces are rectangular, so the figure is a prism.*

*The bases are rectangles, so it is a rectangular prism.*

**B** Identify this figure.

*The surface is curved and there are two circular bases.*

*The figure is a cylinder.*

## Try These Together

1. Is a square a two-dimensional or a three-dimensional figure?
   *HINT: Does a square have the three dimensions of length, width, and height?*
2. How many faces, edges, and vertices are there in the figure of Example A?
   *HINT: Think of a closed box shape.*

## PRACTICE

***Identify each figure.***

3. 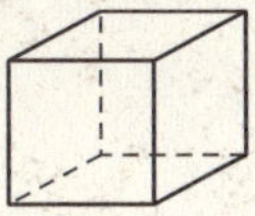

4. 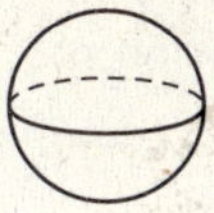

5. 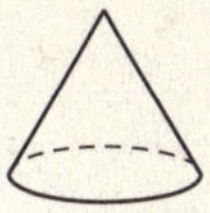

6. How many edges does this rectangular prism have?

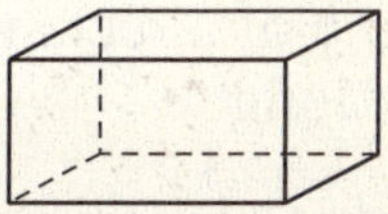

7. **Gift Wrapping** Juanita bought her mother a candle in the shape of a square pyramid for her birthday. How many faces does the candle have for Juanita to cover with wrapping paper?

8. **Standardized Test Practice** How many faces does a triangular pyramid have?
   **A** 4 **B** 3 **C** 5 **D** 2

**Answers: 1.** two-dimensional **2.** 6; 12; 8 **3.** cube (or square prism) **4.** sphere **5.** cone **6.** 12 **7.** 5 **8.** A

NAME ______________________ DATE ____________ PERIOD _____

# 14-5 Volume of Rectangular Prisms (pages 570–573)

The amount of space inside a three-dimensional figure is called its **volume**. Volume is expressed in cubic units.

| **Finding the Volume of a Rectangular Prism** | The volume $V$ of a rectangular prism equals the product of its length $\ell$, its width $w$, and its height $h$. $V = \ell wh$, or $V = Bh$, where $B$ is the area of the base. | 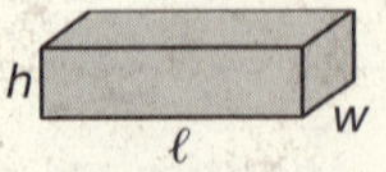  |
|---|---|---|

## EXAMPLES

**A** Find the volume of a rectangular prism that is 8 by 9 by 7 inches.

$V = \ell wh$ *Write the formula.*

$V = 8(9)(7)$ *Substitute the values you know.*

$V = 504 \text{ in}^3$ *Multiply to find the volume.*

**B** A cereal box is 29 cm tall and its top measures 7 cm by 20 cm. Find the volume.

$V = Bh$ *Write the formula.*

$V = 20(7)(29)$ *Substitute the values you know.*

$V = 4{,}060 \text{ cm}^3$ *Multiply to find the volume.*

### Try These Together

**1.** What is the volume of a storage shed 7 feet high with a floor that is 10 feet by 9 feet?
*HINT: Do you know the length, width, and height?*

**2.** A rectangular prism has a height of 2 yards, a width of 0.6 yards, and a length of 1.4 yards. Find the volume.
*HINT: Write the formula and substitute.*

## PRACTICE

***Find the volume of each rectangular prism.***

**3.**
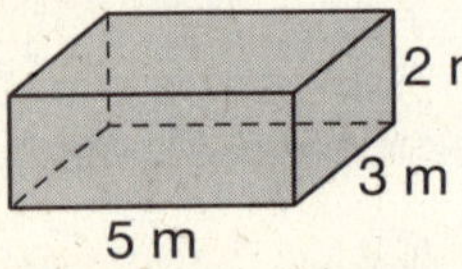


**4.**
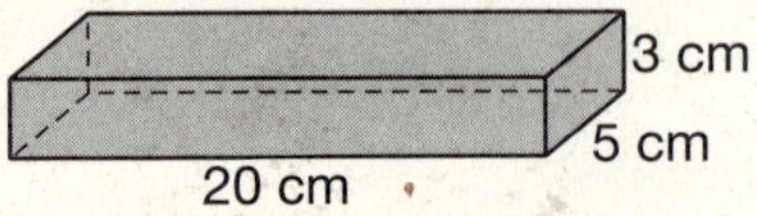


**5.**
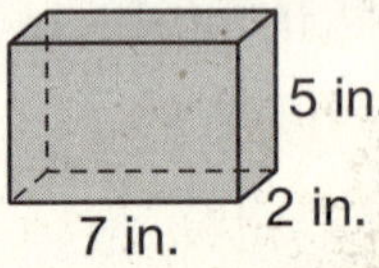


**6.** What is the volume of a rectangular prism that is 12 mm high, 10 mm wide, and 18 mm long?

**7. Hobbies** Mr. Maki is building a new flower bed. The bed is 3 feet wide, 10 feet long, and 1.5 feet deep. How many cubic feet of dirt will he need for his new flower bed?

**8. Standardized Test Practice** Find the volume of a rectangular prism that is 5 feet wide, 8 feet tall, and 11 feet long.

**A** $55 \text{ ft}^3$ **B** $880 \text{ ft}^3$ **C** $440 \text{ ft}^3$ **D** $40 \text{ ft}^3$

**Answers: 1.** $630 \text{ ft}^3$ **2.** $1.68 \text{ yd}^3$ **3.** $30 \text{ m}^3$ **4.** $300 \text{ cm}^3$ **5.** $70 \text{ in}^3$ **6.** $2{,}160 \text{ mm}^3$ **7.** $45 \text{ ft}^3$ **8.** C

NAME ______________________ DATE ____________ PERIOD _____

# 14-6 Surface Area of Rectangular Prisms

(pages 575–578)

The **surface area** of a three-dimensional object is the total area of its faces and curved surfaces.

| **Finding the Surface Area of a Rectangular Prism** | • Find the area of the top and bottom bases.<br>• Find the area of the front and back faces.<br>• Find the area of the right and left sides.<br>Add all these areas to find the total surface area of the prism. |
|---|---|

## EXAMPLES

**A** Find the surface area of a box that is 8 ft by 6 ft by 3 ft.

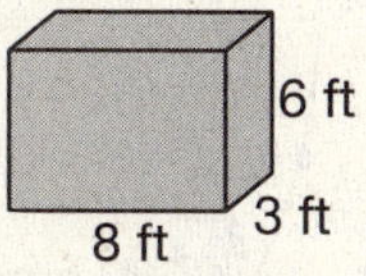


*Area of the top is* $8 \times 3$.
*Area of the front is* $6 \times 8$.
*Area of the side is* $3 \times 6$.
*There are 2 of each face.*
*Total area* $= 2(24) + 2(48) + 2(18)$ *or* $180\ ft^2$

**B** What is the surface area of a rectangular prism with length = 3 in., width = 7 in., and height = 2 in.?

*Area* $= 2(3 \times 7) + 2(3 \times 2) + 2(7 \times 2)$
*Area* $= 2(21 + 6 + 14)$
*Area* $= 2(41)$
*Area* $= 82\ in^2$

### Try These Together

1. Find the surface area of a cube that has an edge of 3 yards.
*HINT: A cube is a rectangular prism with 6 congruent faces.*

2. Find the surface area of a rectangular prism that is 1.3 cm by 2.4 cm by 5.7 cm.
*HINT: Begin by making a sketch and labeling it.*

## PRACTICE

***Find the surface area of each rectangular prism.***

3.
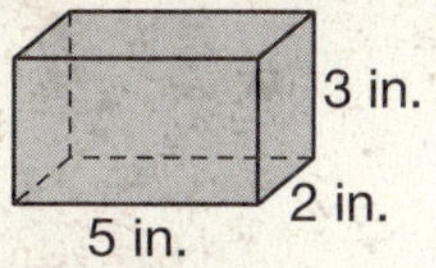


4.
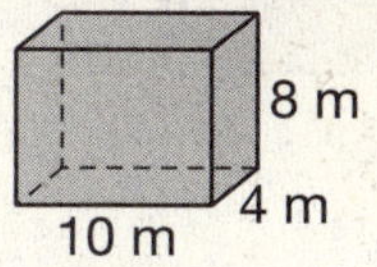


5.
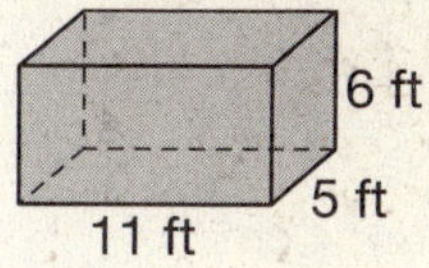


6. length = 12 ft
width = 3 ft
height = 8 ft

7. length = 3 cm
width = 9 cm
height = 1 cm

8. length = 5 m
width = 7 m
height = 8 m

9. **Decorating** Josie is putting wallpaper in her room. If her room is 10 feet wide, 12 feet long and 8 feet high, how much wallpaper will she need? Remember, she will not wallpaper the ceiling or the floor.

10. **Standardized Test Practice** What is the surface area of a 20-cm cube?
**A** 1,200 $cm^2$ **B** 2,400 $cm^2$ **C** 400 $cm^2$ **D** 4,400 $cm^2$

**Answers: 1.** 54 $yd^2$ **2.** 48.42 $cm^2$ **3.** 62 $in^2$ **4.** 304 $m^2$ **5.** 302 $ft^2$ **6.** 312 $ft^2$ **7.** 78 $cm^2$ **8.** 262 $m^2$ **9.** 352 $ft^2$ **10.** B

NAME ______________________ DATE ____________ PERIOD _____

# 14 Chapter 14 Review

## Geometry Carnival

1. You want to make a target like the ones you saw at a carnival. You want the bull's-eye at the center to have less than $\frac{1}{10}$ of the area of the whole target. Does a target with the measurements shown at the right meet this requirement?

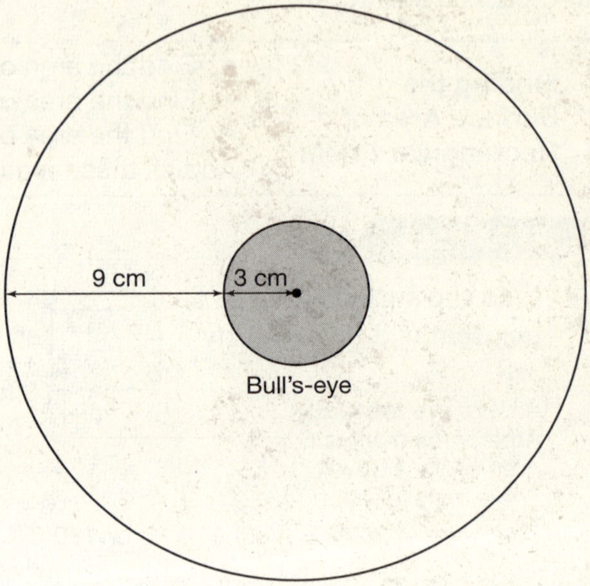


2. At the same carnival, you came across a very interesting game. Two tanks are partially filled with water as shown below. You must place solid prism C into one of the containers without spilling a drop of water to win a prize. Containers A and B are open on the top.

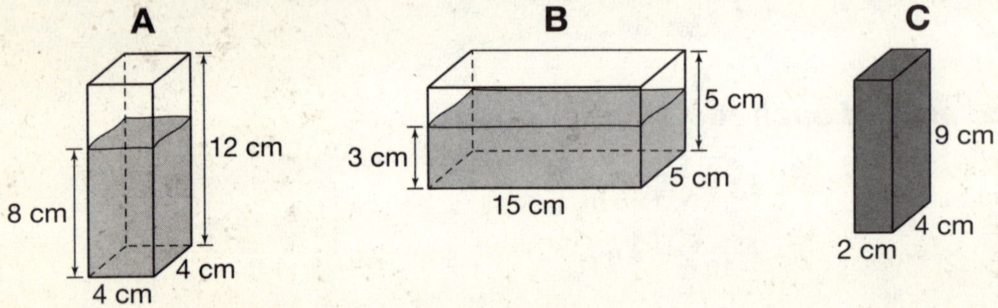


Into which container can you drop prism C without spilling water? Explain.

Answers are located on p. 108.

NAME ______________________ DATE ____________ PERIOD _____

# Answer Key

## Chapter 1 Review

**1.** Plan **2.** 26 **3.** 24 **4.** 2 **5.** 6

When you enter the Middle School Math Club web site, you will gain math power.

## Lesson 2-1

**3.**

| Points | Tally | Frequency |
|---|---|---|
| 10 | 卌 II | 7 |
| 8 | 卌 II | 7 |
| 6 | 卌 | 5 |
| 4 | 卌 | 5 |

## Lesson 2-2

**1.**

**2.**

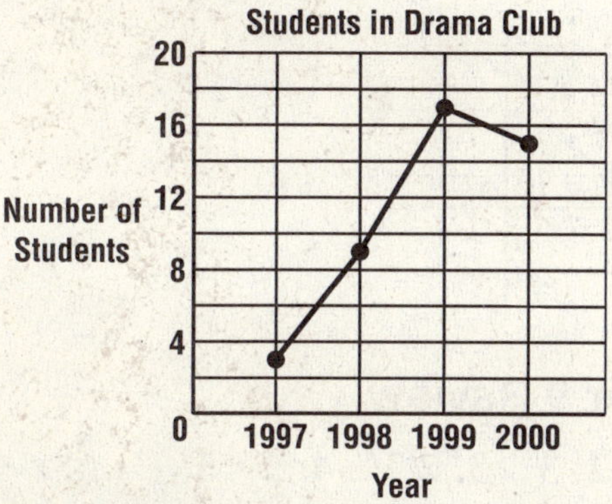


## Lesson 2-5

**5.**

| Stem | Leaf |
|---|---|
| 0 | 2 5 |
| 1 | 4 5 6 |
| 2 | 2 |

*2|2 = 22*

**6.**

| Stem | Leaf |
|---|---|
| 2 | 4 5 |
| 3 | 4 8 |
| 4 | 6 |
| 5 | 8 |

*5|8 = 58*

**7.**

| Stem | Leaf |
|---|---|
| 0 | 2 5 6 8 |
| 1 | 0 2 |

*1|2 = 12*

## Chapter 2 Review

**1.** Stock A, about \$20; stock B, about \$40; stock C, about \$10; stock D, about \$25
**2.** Buy stock B because its value has increased the most over the past several months, and may possibly increase the most over the next several months.

## Chapter 3 Review

**1.** 3, 5, 4, 1, 6, 8, 7, 2 **2.** 0.045 s **3.** 0.15 s **4.** 6.679 s **5.** 52.0 s, 52.0 s, 52.2 s

Working with decimals is fun and easy if you just try a little. Hard work makes easy math for you and me.

## Chapter 4 Review

**1.** 49.28 **2.** 79.28 **3.** 182.344
**4.** 22.793 **5.** 7.35 **6.** 7.35

## Chapter 5 Review

**1.** \$0.80 **2.** The candy bar; it costs \$1.20 and the granola bar costs \$1.25.

**3.**

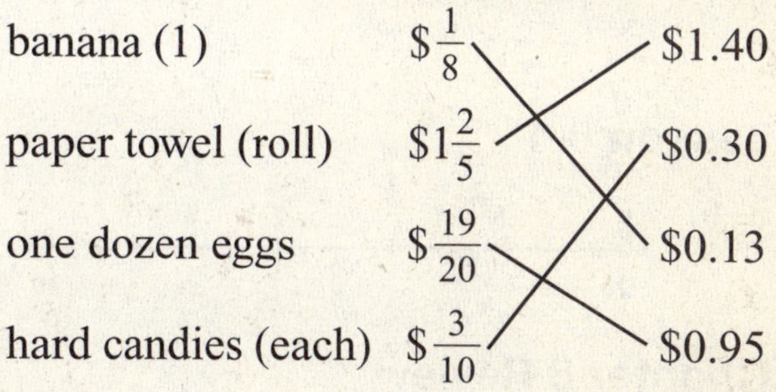


**4.** $\$1\frac{1}{3}$

# Answer Key

## Chapter 6 Review

**Round 1**

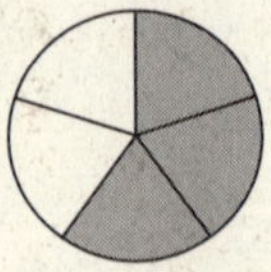

**Round 2**

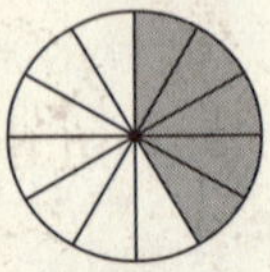

**Round 3**

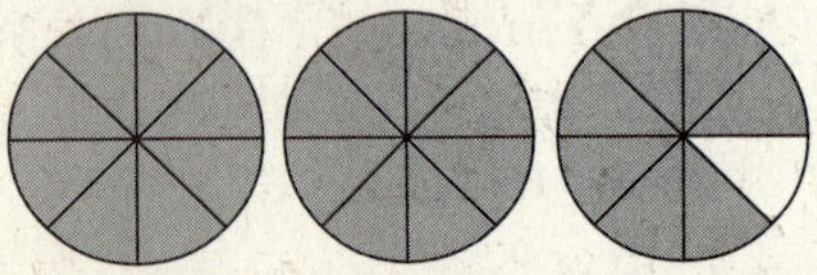

**Round 4**

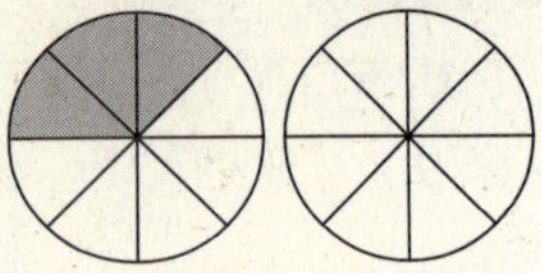

## Chapter 7 Review

**1.**

**2.**

**3.**

## Lesson 8-1

**3–6.**

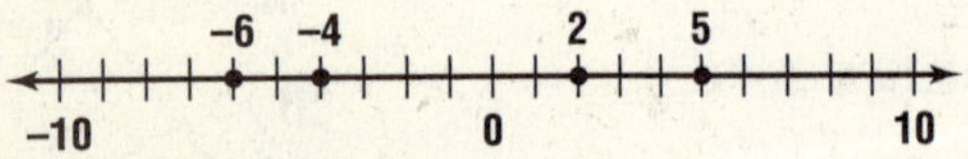


## Chapter 8 Review

**1.** 2 **2.** 6 **3.** 17 **4.** 12 **5.** 16

## Lesson 9-7

**1.**

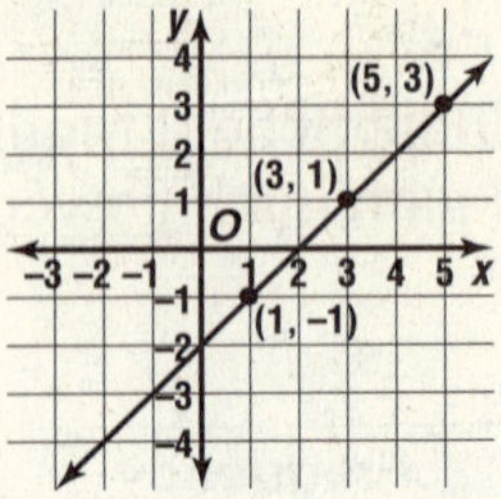


**2.**

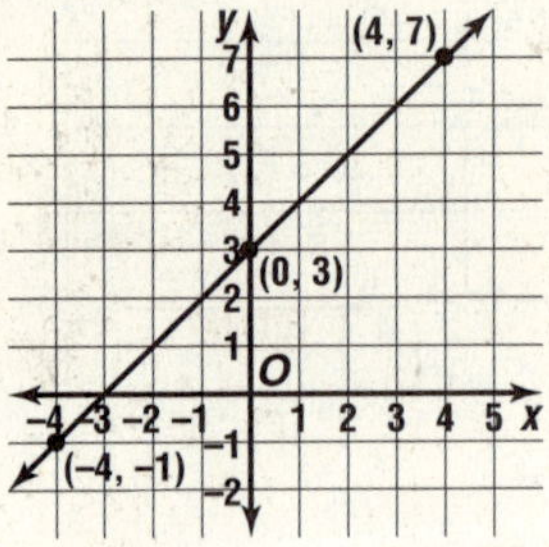


**3.**

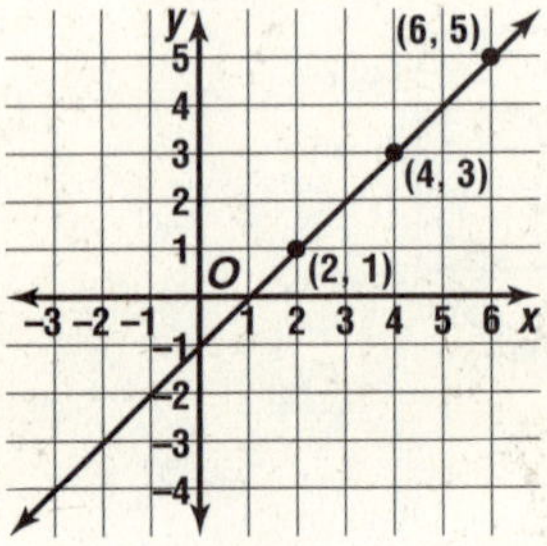


**4.**

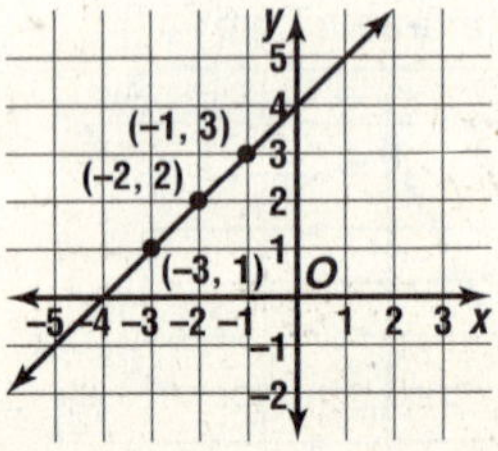


## Chapter 9 Review

**1.** −2 **2.** 3 **3.** 2.3 **4.** 2 **5.** 1 **6.** $x + 4$

NAME ______________________ DATE ____________ PERIOD _____

# Answer Key

## Lesson 10-4

1. 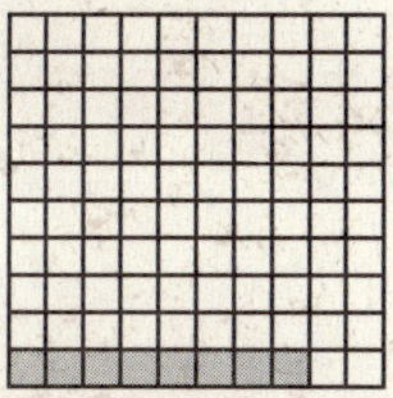

8%

2. 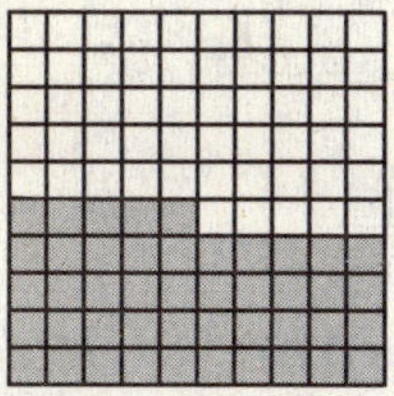

45%

3. 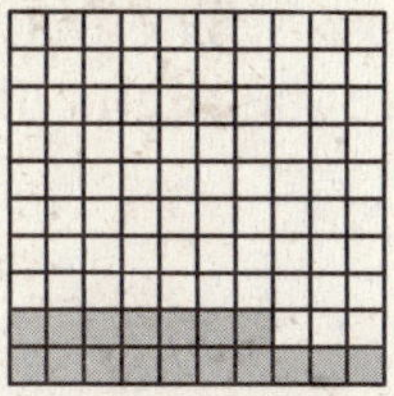

17%

4. 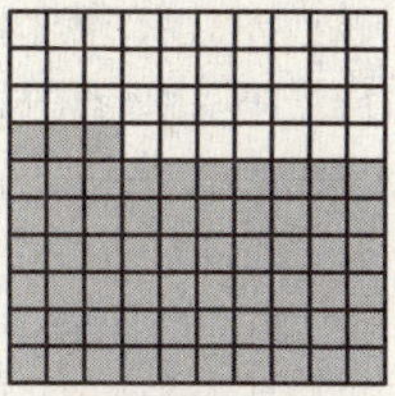

63%

5. 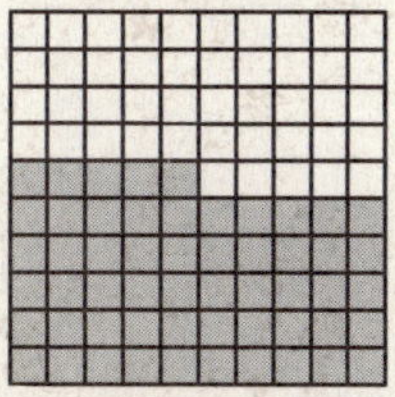

55%

6. 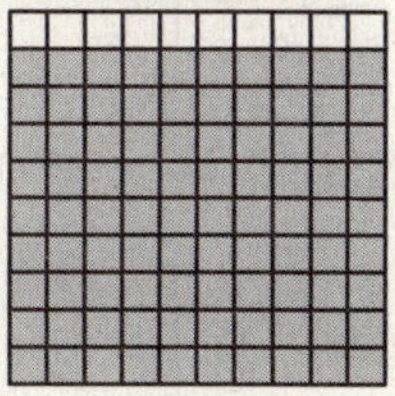

90%

13. 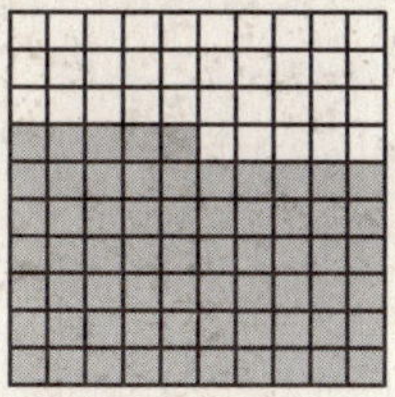

65%

14. 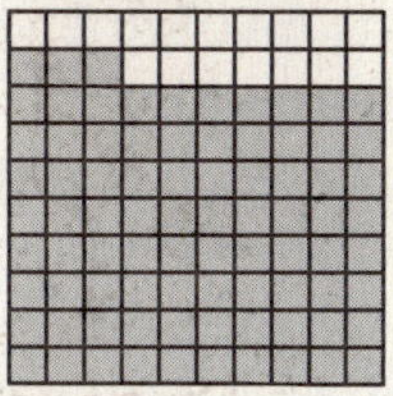

83%

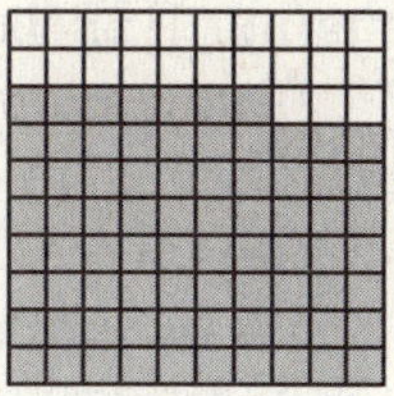

77%

## Chapter 10 Review

**1.** 60 m **2.** 48 m; 36 m

**3.**

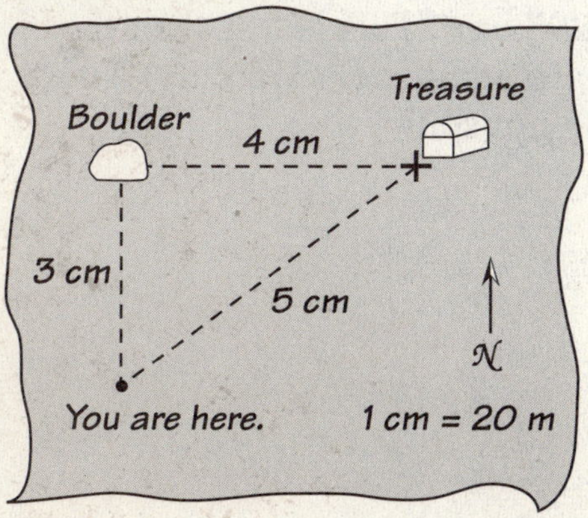


## Chapter 11 Review

**1.** Your family member. To land on the finish square, you would have to roll a 7. This is impossible, so it has probability 0. Your family member needs to roll a 5. This has probability $\frac{1}{6}$. So the probability of your family member landing on the finish square is greater than the probability of you landing on the finish square.

**2.** $\frac{1}{36}$ **3.** $\frac{1}{36}$

## Lesson 12-1

**2.**

**6.**

**7.**

**8.**

## Chapter 12 Review

Answers will vary for measurements.

**1.** 12 in. in 1 ft **2.** 100 cm. in 1 m

**3.** 5,280 ft in 1 mi **4.** $\frac{1}{8}$ ft

**5.** 100 mm in 1 cm; 350 mm

**6.** 0.04 oz in 1 g; 0.16 oz

## Lesson 13-2

**3.**

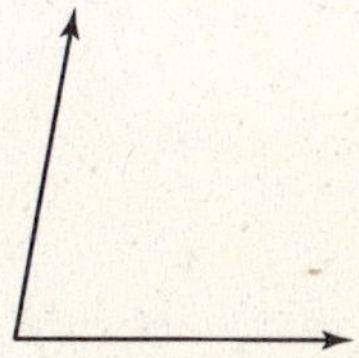

**4.**

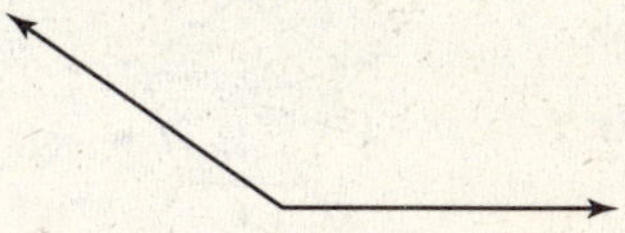

**5.**

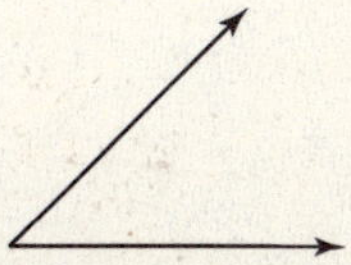

**6.**

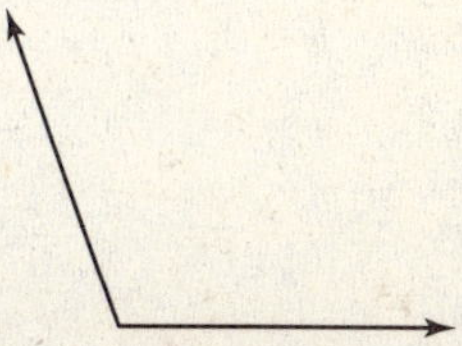

NAME ______________________ DATE ____________ PERIOD _____

# Answer Key

## Lesson 13-3

3.

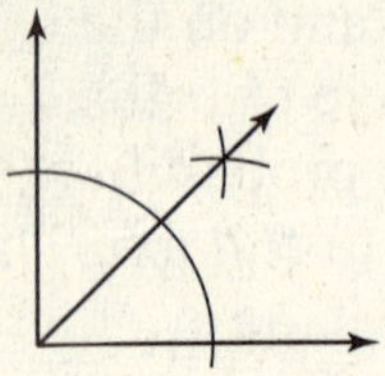

4.

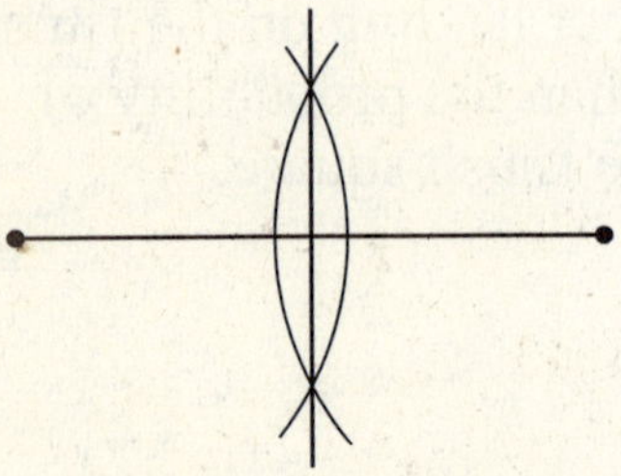

5.

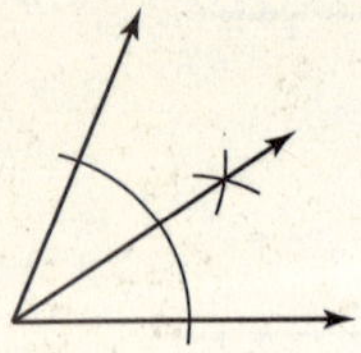

6.

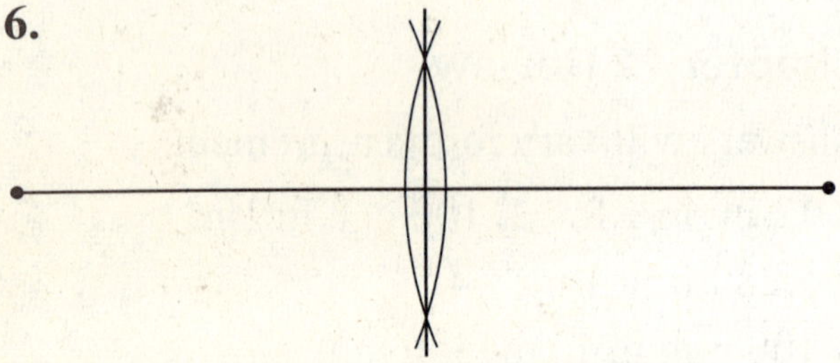

7.

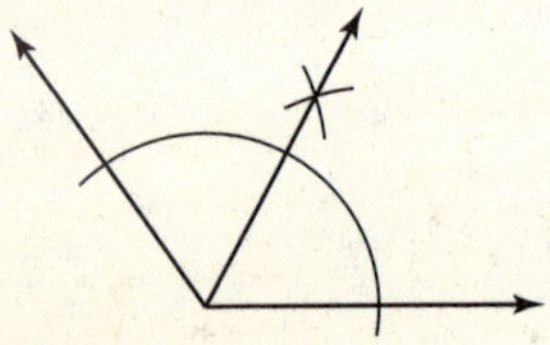

## Lesson 13-5

3.

4.

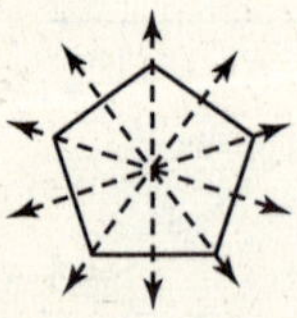

5.

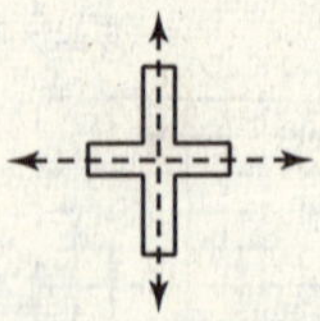

6.

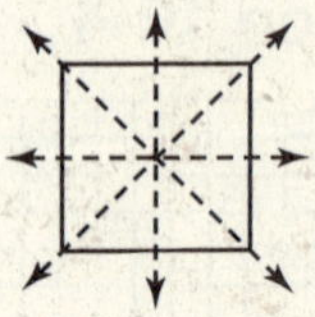

## Chapter 13 Review

**1.** The combination is 36-54-8. The number of sides of the polygons are the digits of the combination.

**2.**

## Chapter 14 Review

**1.** Yes. The area of the bull's-eye is $\pi(3)^2$, or about 28.26 $cm^2$. The area of the whole target is $\pi(12)^2$, or about 452.16 $cm^2$. One tenth of the area of the whole target is about 45.216 $cm^2$, so the area of the bull's-eye is less than $\frac{1}{10}$ of the area of the whole target.

**2.** Container B. Prism C has a volume of 72 $cm^3$. If you add that volume to the volume of water in container A (128 $cm^3$), you get 200 $cm^3$, which is more than the volume of container A (192 $cm^3$). So, placing the prism in container A will cause the water to spill. If you place prism C into container B, which contains 225 $cm^3$ of water, the total volume of the prism and water is 297 $cm^3$. This is less than the volume of container B (375 $cm^3$), so the water will not spill.